Biological Classification

Modern biological classification is based on the system developed by Linnaeus, and interpreted by Darwin as representing the tree of life. But despite its widespread acceptance, the evolutionary interpretation has some problems and limitations. This comprehensive book provides a single resource for understanding all the main philosophical issues and controversies about biological classification. It surveys the history of biological classification from Aristotle to contemporary phylogenetics and shows how modern biological classification has developed and changed over time. Readers will also be able to see how biological classification is in part a consequence of human psychology, language development, and culture. The book will be valuable for student readers and others interested in a range of topics in philosophy and biology.

RICHARD A. RICHARDS is a Professor of philosophy at the University of Alabama. His publications include *The Species Problem* (Cambridge University Press, 2010) and many journal articles.

Cambridge Introductions to Philosophy and Biology

General editor

Michael Ruse, Florida State University

Other titles in the series

Derek Turner, *Paleontology: A Philosophical Introduction*

R. Paul Thompson, *Agro-technology: A Philosophical Introduction*

Michael Ruse, *The Philosophy of Human Evolution*

Paul Griffiths and Karola Stotz, *Genetics and Philosophy: An Introduction*

Biological Classification

A Philosophical Introduction

RICHARD A. RICHARDS

University of Alabama

CAMBRIDGE
UNIVERSITY PRESS

University Printing House, Cambridge CB2 8BS, United Kingdom

One Liberty Plaza, 20th Floor, New York, NY 10006, USA

477 Williamstown Road, Port Melbourne, VIC 3207, Australia

314-321, 3rd Floor, Plot 3, Splendor Forum, Jasola District Centre, New Delhi - 110025, India

79 Anson Road, #06-04/06, Singapore 079906

Cambridge University Press is part of the University of Cambridge.

It furthers the University's mission by disseminating knowledge in the pursuit of education, learning and research at the highest international levels of excellence.

www.cambridge.org
Information on this title: www.cambridge.org/9781107065376

First published 2016

A catalogue record for this publication is available from the British Library

Library of Congress Cataloging in Publication data
Names: Richards, Richard A., author.
Title: Biological classification: a philosophical introduction / Richard A. Richards.
Description: New York: Cambrdige University Press, 2016. |
Series: Cambridge introductions to philosophy and biology |
Includes bibliographical references and index.
Identifiers: LCCN 201601847 | ISBN 9781107065376 (hardback) |
ISBN 9781107687844 (pbk.)
Subjects: LCSH: Biology – Classification – Philosophy.
Classification: LCC QH83.R483 2016 | DDC 578.01/2–dc23
LC record available at https://lccn.loc.gov/2016018473

ISBN 978-1-107-06537-6 Hardback
ISBN 978-1-107-68784-4 Paperback

In memory of my father, Richard Richards

Contents

List of Illustrations

Acknowledgments

Michael Ruse made this book possible. I am indebted to him for his example, encouragement, and generous support. Hilary Gaskin and Rosemary Crawley of Cambridge University Press have been helpful and understanding. I am also indebted to Peter Achinstein. I could not have had a better mentor. My wife, Rita Snyder, has provided support and companionship in dance and life. Robert Olin, the Dean of Arts and Sciences at the University of Alabama, and his Leadership Board have been generous in their support. Scott Hestevold, my chair, colleague, and friend, has also helped make this book possible. I have benefited from conversation and correspondence with David Hull, Michael Ghiselin, Richard Mayden, Mary Winsor, and Matt Haber, and have learned much from the good work of countless others. The views presented here, and any errors, are my own.

Introduction

This is a book about the philosophical history and foundations of modern biological classification. How and why did we come to classify individual organisms into all those Linnaean categories – species, genera, families, orders, classes, and so on? Here *biological classification* will be understood broadly, to refer to the comparison and grouping of organisms, the naming of these groups, the theoretical basis for grouping, and the philosophical foundations for systems of grouping. On this broad construal, it encompasses what is often referred to as 'taxonomy' or 'systematics.'

But why should we care about biological classification? What scientific purpose does it serve? Shouldn't we instead be focused on important biological processes, especially those at the molecular level? This last question was asked early in the twentieth century, as the biologist Ernst Mayr lamented:

> The rise of genetics during the first thirty years of this century had a rather unfortunate effect on the prestige of systematics. The spectacular success of experimental work in unraveling the principles of inheritance and the obvious applicability of these results in explaining evolution have tended to push systematics into the background. There was a tendency among laboratory workers to think rather contemptuously of the museum man, who spent his time counting hairs or drawing bristles, and whose final aim seemed to be merely the correct naming of his specimens. (Mayr 1942, 23)

Today, in the age of genomics, genetics, and epigenetics, we might wonder if these laboratory critics were right. From what we learn about genomes and genes, we can come to understand the molecular basis for development and how this development is affected by environmental factors. And from population genetics, we can learn how these molecular factors change over time, producing evolutionary adaptation and diversification.

Isn't this the future of modern biology, not the old-fashioned, pedantic categorization and naming of organisms based on mere similarities and differences? Perhaps classification is necessary, but it is relatively trivial.

But even on this way of thinking, biological classification is hardly trivial. Surely those who study genomes need to know what kinds of organisms are providing the genomes. In the most obvious case, the human genome project studies just those genomes of members of *Homo sapiens*. And those who study the genomes of primates in general are typically interested in the fact that they are *primates* and share common ancestry. Moreover, to even know that we are referring to the same kind of organisms, we need a stable, precise naming system. For that we need some kind of careful bookkeeping system that associates a particular name with a careful and precise description that allows the identification of individual organisms. To study the genes, genomes, and genotypes of *Panthera tigris*, for instance, we need a way to identify an organism as a tiger – to name it. To study the genes and genomes of bacteria, we need a way to identify an organism as a bacterium. And to study horizontal gene transfer in bacteria, we need to distinguish the different kinds of bacteria. In all these cases, we need a system of biological classification that is precise and unambiguous enough to let us know what kinds of things we are studying.

But the significance of biological classification goes beyond this simple bookkeeping function. Those who study biodiversity are often interested most in a single biological category – species. They may measure biodiversity in terms of species counts, and propose conservation efforts to preserve species. This requires not just that we identify an organism as a particular kind of living thing, but a kind of thing at a particular rank or level – the level of species. As we shall see, the determination of species groupings is far from trivial. There are multiple, conflicting species concepts that divide up biodiversity in different ways, and no obvious reason to prefer one concept over another. The species level has another significance in that *speciation* is taken to be an important evolutionary process. Most basically, it is the appearance of new kinds of living things. But to know when a new species is formed, we need to be able to identify the species of organisms, and we need to know what makes a grouping of organisms a *species* grouping. All this requires some sort of theoretical basis for identifying species and distinguishing the species level from other classificatory levels.

What all this suggests is that the study of many biological processes, even at the molecular level, relies heavily on biological classification, and not just in a trivial, bookkeeping way. But modern biological classification is also practiced within an evolutionary framework and is standardly taken to represent evolutionary history. Organisms are grouped together on the basis of common ancestry, and classification therefore represents patterns of evolutionary diversification. To use a powerful metaphor, it represents the "tree of life." Modern systematists don't simply record the similarities and differences among organisms, "counting hairs and drawing bristles" in Mayr's words quoted above. They are often engaged in the complex and difficult process of reconstructing the evolutionary past. But this process, as we shall see, raises many questions and is fraught with many difficulties. What can we legitimately assume about biological processes in our reconstruction of the tree of life? What is that tree like? And is there even something plausibly described as a tree?

The bottom line is this: Biological classification informs and relies on many other areas in biology. Rather than being an isolated and pedantic activity, it is an activity integrated with, and crucial for the practice of modern biology. Perhaps we can better understand how and why by looking at it from a philosophical perspective.

In Chapter 1, we begin by looking at the cultural and psychological foundations for biological classification. All cultures classify organisms and in broadly similar ways. But this is not surprising, as cognitive psychologists and developmental linguists tell us that hierarchical classification is a natural consequence of our cognitive development and language acquisition. But although it may be natural to classify, not all of our classifications are "natural," in that they pick out important, human independent categories in nature. One way to understand this is through the philosophical framework that distinguishes natural kinds from conventional and artificial kinds.

An important part of understanding modern biological classification is through its history. How did we come to classify organisms this way? Usually the beginnings of modern biological classification are traced to the thinking of Aristotle, in his use of a method of logical division and the classificatory terms 'eidos' and 'genos.' Aristotle has standardly been construed here as an "essentialist," identifying biological taxa with a set of unchanging set of essential generic properties and specific differentia. But as we see in Chapter 2, this understanding of Aristotle's views on classification is, at best, highly misleading. Nonetheless, the Medieval commentators on Aristotle, who were more interested in logic, language, and theology than biology, misunderstood him in similar ways.

In Chapter 3, we continue this historical survey by looking at the beginning of modern biological classification in the medical herbalists of the Renaissance, and the early naturalists, in particular the Swedish botanist Linnaeus. In this period we see the development of a fixed hierarchy and a stable system of naming, along with a turn to theoretical questions. Linnaeus, and other naturalists, including John Ray, Buffon, Kant, MacLeay, and Cuvier began to ask what made a classificatory system "natural," and what it should represent. These questions did not get satisfactory answers until Darwin's evolutionary approach: classification should represent ancestry and the branches on the evolutionary tree.

Darwin's evolutionary framework was adopted and developed in the early twentieth century by Ernst Mayr and G. G. Simpson, who worked to integrate evolutionary classification with new theoretical developments on heredity and natural selection. But by the mid-century, this "evolutionary systematics" found challenges, one from a group of radical empiricists, the "pheneticists," who objected to an evolutionary basis of classification on the grounds that we lack sufficient knowledge of the evolutionary past. A second challenge came from a group of systematists known as "cladists," and was evolutionary in that they took classification to represent the branching of the evolution. But cladists, unlike the evolutionary systematists, rejected the use of assumptions about evolutionary process to reconstruct the past.

Classification has increasingly come to be seen as representing the evolutionary tree, as we see in the various tree of life projects. But there are complications here. First, the Linnaean ranking system is inadequate to represent the many branches of the evolutionary. Second, there seems to be extensive reticulation in the tree of life, with hybridization and horizontal gene transfer. Third, as there are multiple inconsistent trees representing species and genes, it is not obvious that there is a single tree of life. In Chapter 5, we look at these complications, and how we might best think of evolutionary trees.

The species rank seems to be unique, as species are often described as the fundamental units of classification and evolution. But there are multiple, inconsistent ways to conceive species. Chapter 6 looks at this species problem, and the major positions taken about the nature of species. One could be a pluralist about species, recognizing different kinds of species. But some kinds of pluralisms seem to deny the reality of species. One possible solution is that different species concepts function in different ways. However we think about species, there seem to be metaphysical implications, as reflected in the two main views: species as sets and species as individuals. In Chapter 7 we look at how these two ways can be extended to classification at other levels. How should we think about biological taxa at the most general, fundamental level?

Chapter 8 will turn to a philosophical topic that has lurked beneath much of the discussion in the previous chapters: What is the relation between theory and classification? We will begin by looking at the Baconian Ideal, where observation precedes both classification and theory. We will then contrast it with a more theoretical approach, where both observation and classification are theoretically informed. By considering these two approaches, we can better understand the role evolutionary theory has played since Darwin's interpretation of the Linnaean hierarchy. We will finish in Chapter 9 with a look at a fundamental tension revealed in the previous chapters. Our psychology and language learning incline us to think about biological classification as timeless natural kinds with essences. But our best theory tells us instead that biological taxa are historical entities - branches on an evolutionary tree. These two ways of thinking about taxa - as timeless kinds and historical entities - seem inconsistent, and have been implicit in the more than two-thousand-year history of biological classification. Until human psychology or the evolutionary basis changes, biological classification will be fraught with this fundamental tension.

1 Why Classify?

Classification and the Diversity of Life

Nature is filled with a stunning array of living things – animals, insects, plants, fungi, bacteria, and more. This is apparent not just to the biologists who study life, but also to anyone who has ever taken a walk in a park, spent a day at a zoo, watched a nature documentary, or wondered about the pets and pests that share a living space with us. To think about the diversity of life in these terms – as 'animals,' 'insects,' 'plants,' and so on, is to classify it. It seems to imply a division of the world into different *kinds* of things – an animal kind, insect kind, and plant kind.

As anyone who has studied biology knows, modern biological classification goes far beyond the everyday vernacular terms 'animal,' 'insect,' and 'plant,' employing a system based on the ideas of the Swedish botanist Carolus Linnaeus, who developed a framework for classifying living things (as well as minerals) that was hierarchical and comprehensive. According to this approach, all individual organisms are grouped into species that are then grouped together into higher level taxa – genera, orders, classes, and kingdoms. Linnaeus also proposed a naming system based on genus and species membership. He gave humans, for instance, the name *Homo sapiens*, where the first name denotes the genus and the second name identifies the species taxon within the genus. For Linnaeus, an individual human was a member of the species *sapiens*, which was itself a member of the genus *Homo*. *Homo sapiens* was then part of the hierarchy, by being a member of higher level taxa – the order Anthropomorpha, class Quadrupedia, and kingdom Animale.

The Linnaean system was adopted by Charles Darwin and given an evolutionary interpretation. For Darwin, the group-in-group hierarchy of the Linnaean system could represent the branch-on-branch structure of the evolutionary tree, which in turn could represent evolutionary diversification as new species form and diverge. Since Darwin this system has become fully embedded in our practices and institutions – our zoos, natural history museums, biodiversity studies, and collections. Nonetheless, the practice of biological classification is not yet settled. There are four main ongoing philosophical debates about classification. The first is *theoretical*: What should a biological classification represent? Linnaeus may have thought that classification should represent the ideas of God that governed creation, but Darwin and his followers thought biological classification should instead have an evolutionary basis, representing genealogy and degree of divergent change. On Darwin's approach, which came to be known as "evolutionary taxonomy" or "evolutionary systematics," organisms should be grouped together based on common ancestry, but the resulting taxa should be *ranked* on degree of divergence. The Linnaean class Aves, for instance, contains many species of birds, all with a common ancestor. But because this group is so large and has undergone such great modification, it was given an elevated taxonomic rank.

Another evolutionary approach was developed in the second half of the twentieth century by a group of systematists known as "cladists" or "phylogeneticists." They followed Darwin's example in basing classification on genealogy (phylogeny), but rejected the idea that ranking should be based on degree of divergence. According to this approach, a classification should represent only phylogeny, and more specifically, only the branching process in evolution as new species form. This makes a difference in classification. Cladists do not elevate *Aves* to a class, and instead treat birds as the clade (branch) Avialae in Theropoda, a taxon that also includes dinosaurs (Weishampel, Dodson, and Osmólska 2004).

But some systematists have rejected the idea that classification should represent evolutionary history at all. In part this is motivated by the fact that we lack precise knowledge of the phylogenetic origins of all species. We may know, for instance, that all birds share a common ancestry, but we don't know the precise branching order throughout Aves and so can reconstruct only the outline of the evolutionary tree here with any confidence. Consequently we cannot classify on evolutionary grounds with certainty. Moreover, if classification represents evolutionary history and our reconstruction of that history changes, then the classification must change as well. What we can have though, according to these systematists, is a system based on the detailed analysis of similarities and differences among taxa – a "phenetic" system. Organisms that are most similar overall get grouped together at all levels.

The second philosophical debate is about *operational procedures*: How should a classification be generated? The answer to this question is obviously dependent on the answer to the theoretical question about what a classification should represent. Linnaeus primarily used "fructification" traits – traits related to reproduction, thinking that was the best way to uncover the secrets of God's design. The Darwinians who followed, and believed that classification should represent evolutionary history, argued that classifications should be based only on "homologies" – shared traits due to common ancestry, and not on "analogies," similarities based on convergent adaptive change. But how shared traits can be established as homologies has generated some controversy. Evolutionary taxonomists have typically used assumptions about evolutionary processes in general, and the operation of natural selection in particular, to determine which shared traits are homologies and indicate a common ancestry. Cladists (phylogeneticists) have disagreed, arguing that this method is circular. Assumptions about evolutionary processes cannot be used to reconstruct the evolutionary past, because those process hypotheses can be confirmed *only* from the reconstruction of evolutionary history. Cladists have advocated an alternative method based on a parsimony principle they allege to be theory independent. According to this principle, the best hypothesis about the evolutionary past is the one that requires the fewest assumptions of evolutionary change. And in contrast to both the evolutionary taxonomists and the phylogeneticists, those who advocate a phenetic system, based only on similarities and differences, have typically endorsed the use of all similarities and differences, rather than just those traits deemed homologies. This makes sense because the phenetic classification was never intended to represent evolutionary history, only overall similarity.

The third philosophical debate is about the role of *tree thinking* in biological classification. The only diagram in the first edition of Darwin's *On the Origin of Species* was of a branching tree that represented the divergent speciation in evolution. As Darwin used the group-in-group structure of classification to represent the branch-on-branch structure of this tree, the tree metaphor has permeated thinking about evolution and biological classification. Recently there have been attempts to reconstruct the one grand tree of life. But there have also been recent challenges to tree thinking. First, we can construct different and conflicting trees, depending on what interests us: species taxa, organisms, character traits, or genes. The evolutionary history of genes, for instance, has tree-like structures, but gene trees often conflict with species trees. Second, while evolutionary trees typically represent only branching and diverging, there is now believed to be substantial reticulation – the rejoining of branches, through introgression, hybridization, or horizontal gene transfer, especially among plants, bacteria, and viruses. The strictly branching structure of the tree of life does not seem to accurately represent the complicated and messy evolutionary history. Some think we should therefore abandon this tree metaphor.

The fourth philosophical debate, about *ranking*, is a consequence of the idea that classification should be based on the evolutionary tree. *If* biological classification represents a branching evolutionary tree, then the Linnaean hierarchy and naming system appear to be radically inadequate. The current twenty levels or so of the hierarchy cannot possibly represent all the branches of the multibillion-year-old evolutionary tree. How then can we name and organize all the taxa? Indentation and numerical methods have been proposed, but the Linnaean system has become so entrenched in how we think about and represent biodiversity that it is hard to see how it could be abandoned. Is it possible to modify the Linnaean system to better represent evolutionary history and the full diversity of life?

These philosophical debates cannot just be brushed aside. Anyone who is engaged in the classification of living things relies, implicitly at least, on assumptions about what should be represented and how a classification should be constructed. This book aims to look at these issues, not from a partisan perspective (although I have also been a participant in these debates) but from that of a mostly impartial observer. This does not imply that we must avoid any conclusions at all about the various claims, but it does require that we look at them carefully and objectively. But before we look at these issues that have engaged professional systematists about *biological* classification, we need to understand classification in general. After all, modern biological classification is just one species of classification.

What is notable about biological classification is that it need not begin with or depend on the scientific approach based on the Linnaean system. Without consulting biologists we easily distinguish cats from dogs, bees from spiders, birds from fish, and plants from animals. And we seemingly do all this naturally and in the absence of any explicit theory of classification. But why do we classify? One initial answer is that we classify because we must. Classification is an unavoidable natural human tendency. And there is a tendency to classify many kinds of things, not just the living things of our biological classifications. To understand classification in general, we can approach it naturalistically, treating it as natural phenomena to be studied scientifically.

A naturalistic approach reveals that classification is universal. People in all known cultures classify living things, and in roughly similar ways. This is hardly surprising, given what we now know about the psychology of classification. Through observation and experiment, linguists and cognitive psychologists have come to understand what seems to be an innate and universal human tendency to classify in particular ways. In this chapter, we will look first at how people in different cultures think about and classify living things, and then at what developmental linguistics can tell us about the psychological basis for classification. But not all classifications are equal. Some seem to reflect real divisions in the world, while some seem arbitrary or merely pragmatic. On one standard philosophical way of thinking, we can understand this distinction in terms of the differences between *natural kinds* and the merely *conventional* or *artificial kinds*. We will briefly look at this natural kinds framework at the end of the chapter.

A comprehensive understanding of biological classification also requires that we know something about its history. Just as we understand human nature partly though what we know about the evolution of *Homo sapiens* – its origins in a primate lineage and its modification by natural selection and other processes, we can understand biological classification partly through knowledge of its origins and development. In Chapter 2 we look at what many see as the beginning of biological classification in Aristotle's use of the classificatory terms 'eidos' (translated into Latin as 'species') and 'genos' (translated as 'genus'). We will also look at how the Aristotelian framework was adopted and transformed in the 1,500 years after his death. In Chapter 3 we first look at the beginnings of the modern empirical approach to classification in the work of the medical herbalists and early naturalists. Then we delve into how that approach was developed by Linnaeus and given an evolutionary gloss by Darwin. Chapter 4 shows how Darwin's interpretation of the Linnaean framework was further developed in the twentieth century by the evolutionary taxonomists, and then how it was challenged by pheneticists and cladists.

Tree thinking is implicit in the evolutionary approaches that take classification to represent the structure of the evolutionary tree. In Chapter 5 we look at the various ways trees have been used, and the potential problems with trees posed by ranking, hybridization, and horizontal gene transfer. Chapter 6 is on what seems to be the most theoretically significant level in classification – the species level, and the many ways of thinking about species. Chapter 7 is on the metaphysical foundation of classification. How should we think about the basic, fundamental nature of biological taxa? Chapter 8 looks at the relation between evolutionary theory and classification, and contrasts empiricist and theoretical approaches. In Chapter 9 we conclude with what seems to be a fundamental and deep-seated tension between the psychology of classification and the modern scientific and theoretical foundations of biological classification. Our psychology leads us to think about biological classification in one way and our theories about the world lead us in conflicting ways. As this tension lies behind many of the philosophical debates in biological classification, we can perhaps better understand these debates by understanding this tension.

The Anthropology of Classification

One reason to think that classification is natural is that all people seem to do it, not just professional biologists. This is apparent in the studies of *folkbiology* – how the "folk" or nonscientists think about living things, *ethnobiology* and *ethnotaxonomy* – how the members of different cultures think about life and its classification. What these studies seem to reveal are broad cross-cultural similarities in the classification of life. To avoid the bias of modern theoretical biology, ethnobiological studies have typically focused on those cultures least influenced by modern scientific ways of thinking, from the Native American cultures of the Americas, to a variety of relatively isolated cultures of Southeast Asia and Africa. Typically an ethnobiologist will question an educated local "informant" about the names and features of the living things in the local environment, hoping to discover the vernacular terms the informant applies to these, and the implicit classificatory structure. Jared Diamond and K. David Bishop used this method with the Ketengban people of New

Guinea. Over a period of three weeks in 1993, they spent eight to eleven hours a day walking through the forest with their informants, mostly observing birds.

> Our principal method for eliciting bird names consisted in asking Ketengban guides for the name of a bird that we and they both saw, or else heard, while walking together. In order to distinguish which individual bird we meant if there were several in sight or calling, we either pointed to the bird or imitated the call that we were hearing. In order to check that the Ketengban name given in reply actually was meant to refer to the bird about which we were inquiring, we asked our Ketengban guides to describe the bird to us in detail – in particular, its bill, tail, size, color diet, and forest stratum in which it is normally foraged. In that way we could ascertain whether they and we were really talking about the same bird, and whether they were really familiar with the species. (Diamond and Bishop 1999, 23)

Their results:

> We recorded 169 Ketengban bird names, identified most of them definitely, and identified most others tentatively. We also recorded 127 Ketengban names for trees, 51 names for mammals, 34 names for frogs, 16 names for lizards, 9 names for snakes, 6 names for spiders, 4 names for butterflies, and a few names for other insects and fungi, but we will not discuss these other Ketengban names because we do not know the scientific identities of most of them. (Diamond and Bishop 1999, 23)

The last sentence of this quote hints at an obvious complication. Ethnobiologists typically approach and understand the thinking of the local informants within the framework of their own scientifically informed views. If they know much about Linnaean classification, as Diamond and Bishop do, then this likely becomes the basis of the comparison. But Diamond and Bishop also formulated their questioning to uncover ecological or behavioral classifications that would not obviously fit into the modern, evolutionary Linnaean framework:

> Gradually, as we became familiar with many Ketengban names, we structured the questioning by asking our informants to name and describe to us all night birds, or all grassland birds, or all ground-dwelling birds, or all birds similar to some species (e.g., a parrot or pigeon species) whose vernacular name we had already identified. (Diamond and Bishop 1999, 31)

Notice that *Grassland birds* is a hybrid category, based partly on ecology, and groups together birds that are not necessarily closely related in evolutionary terms. Similarly, *night birds* and *ground-dwelling birds* are partly behavioral and ecological groupings. So none of these taxa would fit into the modern evolutionary Linnaean framework based on common ancestry.

The most obvious thing to notice about the findings of Diamond and Bishop is that the Ketengban informants have specific names for, and know a great deal more about the living things in their environment than do the average members of modern, scientific societies. This is unsurprising since their daily life depends much more on knowledge of the animals and plants in their environment than does life in urban and suburban cultures. But another notable conclusion, according to Diamond and Bishop, is that the vernacular Ketengban bird names *seem* to refer to the bird species recognized by scientists, with just a few exceptions where a group of related species might be given a single name, or a sexually dimorphic species might have different names for the female and male (Diamond and Bishop 1999, 35–38).

Diamond and Bishop also looked at the hierarchical structure of the Ketengban classification, and compared it to the Linnaean system:

> Scientific nomenclature for a local biota is hierarchical, with four major levels below the class level (birds being the class Aves). Those four levels are the order, family, genus, and species. In contrast, Ketengban names belong to only two levels: a low-level terminal category corresponding closely to species, and a high-level collective category corresponding approximately to classes or orders. The six collective Ketengban names that we obtained correspond respectively to birds, bats, mammals other than bats, snakes, lizards, and frogs. We found no evidence that Ketengbans name any category intermediate between the low-level terminal category and their high-level collective category. Even though Ketengbans readily understood our questions about naming all species in distinctive bird families, such as naming all parrot species or all hawk species, they offered no name for those intermediate categories (which scientists recognize as families or orders), despite their ability to grasp the bounds of the intermediate category. (Diamond and Bishop 1999, 32)

With the Ketengban there is a group-in-group structure, as in the Linnaean system. The lower level categories are grouped within increasingly higher level categories. And while the Ketengbans have names for only two levels of classification, they seem to recognize intermediate levels between the higher collective level and the lower terminal level.

Studies of the folk classifications of other cultures have arrived at roughly similar conclusions. First, these studies tend to find a relatively large number of names and categories of plants and animals. Brent Berlin's study of the Tzeltal Mayan's classification, for instance, found hundreds of names and categories of plants (Berlin 1999). And Scott Atran found hundreds of recognized categories of snakes, birds, and palms among the Itzaj Mayans (Atran 1999). Second, these studies tend to find a hierarchical structure, usually more complex than what Diamond and Bishop found in the Ketengbans. According to Berlin, there are up to six levels or ranks in the hierarchies he has studied. The highest is *kingdom*, followed by *life form*, *intermediate*, *generic*, *specific*, and *varietal* (Berlin 1992, 22). Not all of these hierarchical levels would necessarily be found in all cultures though. Nor will all the levels necessarily have a name (Berlin 1992, 31–33). Ralph Bulmer, in his studies of Kalam ethnotaxonomy, identified five levels of hierarchy, from the highest, *primary taxa*, which are not subsumable into any larger taxon, to the *terminal taxa*, which have no named subdivisions (Berlin 1992, 65). Third, there also seems to be a basic privileged level in the hierarchy. According to Berlin this is the *generic* level. Most of the taxa are found here. These taxa are easily recognizable and have simple names (such as 'dog' and 'cat' in the vernacular English; Berlin 1992, 64). Atran dubbed the groupings at this level 'generic species' (Atran 1999, 124).

What generalizations can we draw from these studies in the folkbiology of various cultures? First, all folk classifications seem to be hierarchical, with a group-in-group structure. Although Diamond and Bishop found only two explicitly named levels in their investigation of Ketengban folk classification, more levels were implicitly recognized. Berlin, Bulmer, and Atran have all found more, albeit not always named, levels. Atran argued that there is a default hierarchy comprising five levels. According to Atran, the highest level is *kingdom* (Atran 1999, 122). Folk kingdoms, if not explicitly named, may be indicated by the use of a particular suffix or term. The Itzaj used a particular term only for plants, for instance. They also used a term translated as 'forest-thing' at the kingdom level that includes many vertebrates, invertebrates, birds, and fish (Atran 1999, 123). The next highest default level is *life form*, a level that is often based on functional, developmental, or ecological factors:

> The majority of taxa of lesser rank fall under one or another life form. Most life-form taxa are named by lexically unanalyzable names (primary lexemes), and they have further subdivisions such as tree or bird. Biologically, members of a single life form are diverse ... Life-form taxa may represent general adaptations to broad sets of ecological conditions, such as the competition of single-stem plants for sunlight and tetrapod adaptation to life in the air. (Atran 1999, 122)

Among the Itzaj Maya, life forms in the plant kingdom include *trees*, *shrubs*, *vines*, and *grasses*. Animal life forms include categories roughly corresponding to mammals (excluding bats), birds (with bats), and herpetofauna (amphibians and reptiles) (Atran 1999, 123, fn. 5).

The third highest level is the *generic species*, usually identified as species-like groupings. This level seems to have a special status, and for a variety of reasons, linguistic, inferential, psychological, and developmental:

> The rank of generic species is the level at which morphological, behavioral and ecological relationships between organisms maximally covary. The majority of Itzaj folkbiological taxa belong to this level. It is this level that Itzaj privilege when they see and talk about biological continuities. Generic species represent cuts in nature that Itzaj children first name and form an image of ... and that Itzaj adults most frequently use in speech, most easily recall in memory, and most readily communicate to others ... It is the rank at which Itzaj, like other folk around the world, are most likely to attribute biological properties, including characteristic patterns of inheritance, growth, physiological function, as well as more "hidden" properties such as hitherto unknown organic processes, organs and diseases. (Atran 1999, 127)

The level of *generic species* seems to be the primary focus of how the Itzaj and other folk talk and think about living things. It is generic species that are most easy to identify, name, remember, and think about. And it is at this level that there is the strongest tendency to generalize: from the fact that some individuals of a generic species have some trait, other individuals must also have that trait. Generic species also typically have simple names, such as we see with the English vernacular terms 'dog,' 'cat,' 'oak' and 'robin.'

The lowest levels are the *folkspecific* and *folkvarietal*. Some of the most culturally significant generic species get subdivided into folkspecifics that are distinguished in part by their names. Generic species tend to have simple names, whereas folkspecifics tend to have binomials constructed out of the generic species name, such as we see with 'white oak' and 'mountain robin.' At the lowest level is the *folkvarietal*. This level is often indicated by a trinomial, as in 'spotted white oak' versus the 'swamp white oak,' and 'yellow village papaya' as distinguished from the 'white village papaya.' The Itzaj typically distinguished folkvarietals among those living things that have significant perceptual differences (color perhaps) and specific practical uses or cultural significance (Atran 1999, 127–128).

Many who study these folk classifications claim there is a standard folk level of classification corresponding to the modern scientific species level. Atran claims, for instance: "Humans everywhere classify animals and plants into species like groupings that are as obvious to a modern scientist as to Maya Indian" (Atran 1999, 120). And Diamond and Bishop, as quoted earlier, claim that for the Ketengban there is a "low-level terminal category corresponding closely to species" (Diamond and Bishop 1999, 32). Diamond and Bishop argue further that this correspondence is evidence for realism about species taxa.

> If peoples with very different upbringings and motivations for naming nevertheless tended to recognize the same units of Nature, that would lend support to the view that those units correspond to an objective reality that is not a mere invention of scientists. (Diamond and Bishop 1999, 17)

If people all over the world and in different cultures divide up biodiversity into species-like groupings in the same way, then there must be some objective, human independent significance to the units of grouping and division. There really must be these species-like groupings in nature!

But there are problems here. First, as we shall see in Chapter 6, there is substantial disagreement among modern systematists about the nature of species, and consequently how species taxa are identified and individuated. Some systematists appeal to reproductive cohesion and isolation, others to morphological, behavior or genetic similarity, and yet others to the fact that species are lineages. This leads to disagreement about species counts – the number of species taxa represented by a group of organisms. If so, there is no single set of species taxa recognized by all scientifically trained systematists, and we should be skeptical of the claim that folk classifications recognize precisely the same species taxa every modern systematist recognizes. The claim that the folk recognize the same species as modern scientists wrongly assumes agreement among scientists.

Second, many of the scientific species concepts are "dimensional" in the sense that they conceive of species taxa extending in space as populations, and over time as lineages. And within the populations and lineages, scientific thinking about species allows for substantial variation. This is a consequence of the fact that scientific thinking about species taxa is informed by evolutionary theory. But just because a culture names a kind of living thing that is part of a population and lineage does not imply that the members of the culture think of that kind of living thing in these ways – dimensionally as populations and lineages. The naming of a living organism as a kind of thing is compatible with many different conceptions of the *kind* itself. The folk might be thinking of the species-like category in a variety of ways, from various kinds of similarity, but also perhaps on religious or mystical grounds, in ways not shared with modern systematists. The folk informant and the ethnobiologist may therefore agree that a small set of organisms may represent the same number of groupings, but have different ideas about what that means in terms of the theoretical basis of the category or overall hierarchy. These differences may affect how new specimens might get classified. One cannot therefore conclude from the same limited number of groupings that the classifications are the same.

Third, these species-like groupings are not necessarily at the species level of the Linnaean classification, as Atran acknowledges:

> Generic species often correspond to scientific genera or species, at least for those organisms that humans mostly readily perceive, such as large vertebrates and flowering plants. On occasion, generic species correspond to local fragments of biological families (e.g., vulture), orders (e.g., bat), and especially with invertebrates, higher-order taxa. (Atran 1999, 125)

The Linnaean classificatory levels don't get distinguished in folk biology because, as Atran points out, many of the genera in any particular place are monospecific – represented by only a single species. If so, there would be no reason to distinguish the species level from a higher level in the classification. A single folk grouping may then be ambiguous between species and genus levels. Moreover, when there are multiple species of a single genus, these species may be difficult to distinguish perceptually. There may be too little of a morphological gap for the folk taxonomists to distinguish the two species (Atran 1999, 125–126).

Lurking behind these three problems is a more general philosophical problem – how to determine the *mapping* of a term in one classificatory system onto another system. When a proper name is applied to an individual object, the mapping of the name is relatively unproblematic. A particular horse can be dubbed 'Secretariat' and the reference of the name is clear – the name refers to a particular horse, but not to any other horse, not to any other creature, nor to any other object. But the mapping of a *general* term such as 'horse' to things in the world is more problematic. We can apply the term to a particular individual horse, perhaps the one named 'Secretariat,' but it isn't necessarily clear to what other objects the term 'horse' would apply.

This problem can perhaps be best understood through an example. Suppose a culture has experience and knowledge of only a single species (in the modern scientific sense) of horse, *Equus caballus*, and the members of the culture use the term 'horse' on the basis of this experience to refer to particular horses they experience and that are of this species. But why should we interpret the use of the term 'horse' to refer to *just* living things in the scientific species taxon *Equus caballus* rather than living things in the genus *Equus* or family Equidae? The individual horses are just as much members of the genus and family as they are the species. In other words, why should we take the vernacular term 'horse' to refer just to things of a particular species, and therefore at the species level? The people who use the vernacular term do not make any of the distinctions between the various species of the genus *Equus* or the genera of the family Equidae. Whether they could and would make the distinction between things of different species, and what terms they would use is an open question.

There are at least two options here. First, we could claim that 'horse' maps onto one of the three taxanomic levels – *Equus caballus, Equus*, or Equidae. If so, then we would need to give reasons why one mapping is better than other. Moreover, we would need to give a reason why it shouldn't map at an even higher level. Second, we could just conclude that the local vernacular term 'horse' is indeterminate relative to the Linnaean system, and cannot be identified with a particular taxon or classificatory level.

In short, there is no determinate mapping *as the term is used*. Now it may be that when confronted with members of different species in *Equus* or Equidae, these people would come to use the term 'horse' differently, perhaps with some linguistic indicator to distinguish the different kinds of 'horse.' But that would be a different classification, one that recognized different levels in a hierarchy of 'horses.' That would be a different use of the term! The point is this: Ethnobiologists cannot just assume a word gets mapped to either a particular group of organisms (a species taxon) or to a category level of organisms (the species category). Whether it does so depends on the relation between the two systems as they are used by the respective linguistic communities. Do they make the same distinctions and have the same categories and levels of hierarchy? If not, it is hard to see how the folk taxonomies can be said to recognize precisely the same set of species recognized by modern biologists. The most we could say is that *within a narrowly prescribed context*, some of the folk taxonomic terms are coextensive with some of the scientific, Linnaean terms.

An additional complication in the comparison of folk and scientific classifications is that many folk categories are utilitarian, in the sense that they are based on practical value. Those things that serve important cultural and practical roles are more likely to be classified, and the resulting classifications often reflect these cultural and practical differences. The Sahaptin Indians of Washington and Oregon, for instance, eat fish but not birds, and have much richer classifications for fish than for birds (Hunn 1982, 834). Similarly, the Tzeltal Indians have spent little effort in the classification of adult butterflies (*Lepidoptera*), but much effort in classifying the larvae, grouping them according to whether they are edible or attack crops. Similarly, their folk botanical categories are based on whether plants are poisonous, invasive, or edible (Hunn 1982, 831). Among the Itzaj Maya, snakes that are deadly are classified separately from those that are not (Atran 1999, 161). Animals and plants that fit into a religious or mystical narrative are sometimes classified differently than those that do not. The Karam of New Guinea, according to Ralph Bulmer, don't classify the cassowary as a bird, because of their views about its spiritual significance (Bulmer 1967).

There is little dispute among ethnobiologists and ethnobotanists that folk classifications of various cultures sometimes reflect the different utilities, roles, and functions of the various kinds of living things in each culture. But there is disagreement about the relative significance of utilitarian criteria in biological classifications. The "utilitarians," such as Eugene Hunn, claim that folk taxonomies are fundamentally based on practical criteria. Hunn adopts this view in part because he thinks we should understand classification adaptively, within an evolutionary framework. Humans, like other creatures, classify living nature based on utility in survival and reproduction:

> Vervet monkey alarm calls, for example, differentiate among leopards, martial eagles, baboons and pythons as predators and elicit distinct evasive responses in wild monkeys ... Lions are known to exhibit a variety of hunting strategies as a function of the prey species. If they did not, they would certainly be clumsy hunters ... Our presumably innate propensity to "see" biologically natural categories allows our behavior to be flexible in a highly efficient manner. (Hunn 1982, 833)

This focus on the practical might be expected to sometimes result in the categorization of what modern systematics thinks of as species, because of patterns of similarities among the members of species. Members of a single species in the modern sense will likely have the same utility because of similar morphologies, behaviors, and so forth. But such a classification will also be influenced at other levels by utility – whether a kind of animal or plant is dangerous, edible, or poisonous. Categories of snakes above the modern species level might be determined not by evolutionary relationship, as in the modern biological system, but by the risk they pose to humans.

The "intellectualists," on the other hand, argue that the fundamental criteria are instead based on perceptual factors and structural features of nature. For the intellectualists, the utility of a category may have *some* relevance to its place in the classificatory framework, but the primary determinants are perceptual and structural (Medin and Atran 1999). Classifications, according to the intellectualists, are therefore more objective in that they do not depend so much on subjective factors – the different preferences and practices of various cultures. However this debate between the intellectualists and utilitarians might be resolved, the merely utilitarian categories in folk classifications distinguish them from the modern scientific classification. The category *weeds*, for instance, is widely recognized as a utilitarian category in cultures that also use the scientific Linnaean system, but it is not part of that classificatory system. It seems that folk taxonomies may be eclectic, incorporating practical and cultural criteria as well as the biological, in ways that the modern evolutionary Linnaean classification is not.

So what can we conclude from these ethnotaxonomic studies? I suggested earlier that biological classification is a natural human tendency and we do it because we must. The studies cited here suggest this claim is at least plausible - even if folk classifications differ from modern biological classifications in important ways. At a minimum, all cultures seem to classify living things into hierarchies. But there is clearly more to a naturalistic understanding to classification. As we noted at the beginning of this chapter, language itself seems to imply classification. The use of terms like 'horse' and 'dog' implies a classification distinguishing a horse kind and a dog kind. And when we have binomials, such as 'red oak' and 'white papaya,' we get an implicit hierarchy. *Red oak* is a subdivision of *oak* and *white papaya* is a subdivision of *papaya.* If so, the universal nature of classification is hardly surprising, given that language itself is universal among human cultures. What this suggests is that the similarities among folk classifications might be due, not just to the recognition of the real units of nature, but also to similarities among people. It may be that all cultures classify because virtually all humans have the same psychological foundation for classification. In the next section we will look at this possibility.

The Psychology of Classification

We can begin to understand the psychology of classification as many psychologists do, through language. The mere use of terms such as 'cat' and 'dog' seems to imply a classification. There are at least two different kinds of things, a *cat* kind and a *dog* kind. And when we learn to use these terms, as well as many others, such as 'insect,' 'animal,' 'plant,' and so forth, we must learn how to map these terms onto things in the world. We need to learn what things the terms legitimately apply to, and what they don't. In other words, we need to learn the *extensions* of these terms.

This way of thinking about classification, as a function of language, can be applied to ethnotaxonomy. After all, the members of each culture must learn how to map local terms onto the world. But while the study of folk classifications can reveal the similarities and differences across cultures, it cannot by itself tell us everything we need to know, in particular about the *mechanisms* underlying the linguistic basis of classification. For an understanding of the mechanisms perhaps we can turn to developmental psychology, as argued by psychologist Sandra Waxman:

> For no matter how carefully an ethnobiological record is constructed and analyzed, or how elegantly an experiment is designed, evidence from adults cannot reveal the origins of knowledge or the mechanisms responsible for its unfolding. It is impossible to discern the initial state of a system from an examination of its mature state. To understand the origins and emergence of a system, one must begin at the beginning. (Waxman 1999, 239)

The "beginning" here is the development of classification in each person. This process is implicit in the development of language.

By nine to twelve months of age, infants have typically begun to learn and produce words. By twenty-four months, they can normally produce hundreds of words and put them together into simple, well-formed phrases. But infants do not learn words indiscriminately. In the beginning they seem to learn a disproportionate number of nouns. One study of the first fifty words learned by infants showed that about half of all the earliest acquired words are *general nominals*, nouns that refer to all members of a particular category – classes of objects, animals, people, letters, numbers, and so forth. The next most common type of words learned were *specific nominals*, words that refer to specific instances of a category. This linguistic category includes proper names, but also nouns that refer to individual things, such as 'mommy,' a single instance of the class *mother*. *Action words*, which describe or demand an action, are learned at about the same rate as specific nominals. Much less common among these first fifty words are *modifers* that refer to properties or qualities of things, *personal-social words* that express affective states, and *function words* that serve grammatical functions (Capone et al. 2010, 202–203; Nelson 1973). But for *general nominals* an object category must be learned along with the *object name*. As an infant or child learns the word 'horse,' he or she also learns *horse* as a category, and presumably something about that category, insofar as that term can be applied to things other than the original dubbed object. This process is far from trivial. We don't see and cannot point to the category *horse* in the same way we can see and point to an individual object when learning a term.

Nor do we, or can we, learn a category by observing *all* members of that category – all the horses that exist, have existed, or will exist. What is impressive is that infants learn object categorization easily, but without perceptual acquaintance with anything like the full contents of the category.

But even what may seem straightforward here is not. Learning an object name is far from trivial, as the developmental psychologist Ellen Markman explains:

> When a child hears a word used to label an object, for example, an indefinite number of interpretations are possible for that word. The child could think that the speaker is labeling the object itself as a whole, or one of its parts, or its substance, or its color, texture size, shape, position in space, and on and on. Given the impossibility of ruling out every logically possible hypothesis, how is it that children succeed in figuring out the correct meaning of terms? (Markman 1989, 8)

In some dubbing ceremony, whereby a word is applied to a new object, the speaker *might* be pointing to any number of things: an object, a part of an object, a property of the object, the size of the object, the stuff the object is made of, the position of the object, what the object is doing, or even one's attitude toward the object. We might, for instance, point to something and utter the words 'tasty' or 'yuck.'

The problem is worse for object categories such as *horse*. Not only must an infant understand that a word applies to a particular whole object, but that it also applies to other things of a particular kind. But what kind of kind is not determined by the dubbing. One could point at something, dub it a kind, but that kind could be based on parts (having four legs), size (being larger than humans), properties (being brown), location (being in the ocean), or action (flying), and what attitudes we have toward the objects (being tasty or desirable). Sandra Waxman explains:

> We know that the perceptual and conceptual repertoires of infants and young children permit them to appreciate many different kinds of properties of objects and relations among them. In principle, infants' rich and flexible repertoires should complicate the task of mapping a word to its meaning. How do infants select among the various kinds of properties and relations when seeking to determine the intension and extension of a word? How do they so rapidly learn that a given word (e.g., *tapir*) will apply

> to a particular whole object and can be extended to other members of its kind ... but not to salient parts or properties of the object (e.g., its long snout or lackluster color), to salient actions in which it may be engaged (e.g., foraging in the ferns), or to other salient thematic or associative relations involving the named object? If infants had to rule out these (and countless other) candidate meanings, word learning would be a laborious task, and would proceed at a sluggish pace. Yet this description does not fit. Infants and toddlers acquire words, especially words for objects and object categories, at a remarkable pace. (Waxman 1999, 244)

As suggested at the end of this passage, one way to solve this reference problem might be to consider all possible hypotheses and eliminate the incorrect ones. To do this an infant would first, need to consider whether a word referred to a whole object; some particular part of an object; some particular property of an object; something the object was doing or had done to it; or a relation or subjective response to the object. An infant would then need to consider ways to extend the term to other things, whether those things be objects, parts of objects, actions, relations, or subjective responses. But this complex analytic process is clearly beyond the cognitive abilities of infants and children (and likely of most adults as well). Moreover, infants and children learn words far too quickly and easily even if they were to have these abilities. If so, how do they actually learn object names and object categories?

The answer, according to Markman, is that many of the alternative hypotheses are not even considered (Markman 1989, 8). We can think of this as a sort of linguistic bias, a tendency to interpret terms in some ways and not in others. Sandra Waxman describes this bias toward some alternatives as 'expectations' (Waxman 1999, 243). Capone et al. use the terms 'bias,' 'constraint,' and 'principle' (Capone et al. 2010, 206). Whichever term is most appropriate to describe these tendencies in language learning, there is agreement that, first, they operate in normal language development; second, they affect how we categorize things in the world; and third, they are not learned in an ordinary way. Sandra Waxman explains:

> It is unlikely that this initial expectation could have been learned or induced by the infant on the basis of observations for their existing word mappings ... for very few such mappings (if any) have been established by 9 months of age. From the outset, then, novel words direct infants' attention to object categories. The power of the dubbing ceremony derives, in large

> part, from the infants' a priori expectations that novel words, applied to individual objects, will refer to those objects and to other members of the same category. This initial rudimentary linkage between words and object categories serves a crucial function: it guides infants in their earliest efforts to map words to categories of objects. (Waxman 1999, 259)

We can divide these language biases into four main clusters. The first cluster is related to the application of words to things in the world. Here the first and most obvious principle is a *whole object bias*. When an infant or child is presented with a novel word, whether particular or general, it is taken as a rule to refer to the whole object rather than a part or feature of that object (Capone et al. 2010, 206; Waxman 1999, 244). The second principle in this cluster, the *principle of exclusivity*, asserts that only one name is allowed for each object. So if another novel term is introduced, that term is normally taken to apply to another, previously unnamed object. Finally, according to the *principle of conventionality*, names don't change. If an object has been dubbed with a particular word, that word will continue to apply. The value of this principle should be obvious. If names were allowed to change regularly, there would be little value in learning an object name (Capone et al. 2010, 206–7).

The second cluster of biases is related to the formation of object categories. The *principle of extendibility* is basic here. According to this principle, a word is typically taken to refer not just to single objects, but also to objects that are similar in various ways. So when confronted with a novel term, an infant will automatically try to extend that term to other objects. Because of a *shape bias*, the extension of a term will be typically extended based on perceived shape. We can readily see the shape bias at work in some of children's naming errors. For example, a child may look at the moon and say *ball* because both the moon and ball are round. The shape bias, as understood here, is based not only on such obvious global characteristics as being round, but also on presence of eyes for living things, or having wheels for cars. Size is apparently not a bias here, as a child will readily apply the term 'car' to both a toy car and a real car. Terms are extended as well in terms of function. Infants and children will often use the same word for two objects with different shapes if they have the same function (Capone et al. 2010, 207–208). Based on this *function bias*, the word 'chair,' for instance, could be extended to new objects of different shapes based on seating function.

The third cluster of biases is related to the development of hierarchies. In the earliest period of language development, infants do not distinguish nouns from other kinds of words, and will treat adjectives as object names and categories. But at around twenty-four months, they start to distinguish words based on different syntactic functions, distinguishing nouns from adjectives and other kinds of words. The use of adjectives to modify nouns seems to stimulate the development of hierarchies. When an infant learns a noun such as 'oak,' he or she also learns an object category. But when an adjective–noun phrase such as 'red oak' is learned, it seems to generate a subcategory within the 'oak' category. So as infants learn complex object names and categories, they also acquire classificatory hierarchies – categories within categories (Waxman 1999, 251).

Within these hierarchical classifications, not all levels are equal. Cognitive psychologists, like the ethnobiologists, identify a *basic level* that is privileged, with superordinate levels above and subordinate levels below. The basic level seems to have the following features. First, it is the most inclusive level, where members possess numerous common attributes. Psychologists claim that this basic level similarity supports inductive generalization. We can more reliably generalize from the members of basic level categories to other members than we can in higher level categories. And wider generalization is possible at the basic level than at lower level categories. Second, the basic level is the level first learned by children, and the level most quickly identified by adults. Unsurprisingly, it is also the level most likely to be named (Medin and Waxman 1998, 169). But, it should also be noted, there is some variation in what counts as the basic level, in part based on familiarity and expertise, which varies across cultures (Waxman 1999, 270). Whatever the case, it seems that classifications in infants and children seem to be built from the center out – from a basic level upward to superordinate levels and downward to subordinate levels.

The fourth cluster of biases is related to how infants and children think about the nature of categories and the objects in the categories. Some have argued that infants and children are born essentialists, in that they have certain psychological tendencies. Susan Gelman has made this claim:

> I contend that essentialism is a pervasive, persistent reasoning bias that affects human categorization in profound ways. It is deeply ingrained in our conceptual systems, emerging at a very young age across highly varied cultural contexts. Our essentializing bias is not directly taught, nor does it

> simply reduce to direct reading of cues that are "out there" in the world. Most decidedly, it is neither a late achievement nor a sophisticated one. (Gelman 2003, 6)

There is even a bias in how this bias is applied:

> The question of which categories we essentialize is a tricky one. In a nutshell, I argue that essentialism is the result of several converging psychological capacities, each of which is domain-general yet invoked differently in different domains. Collectively, when these capacities come together to form essentialism, they apply most powerfully to natural kinds (including animal and plant species, and natural substances such as water and gold) and social kinds (including race and gender), but not to artifacts made by people (such as tables and socks). (Gelman 2003, 6)

Gelman argues that this psychological essentialism has three main components. First, people believe that certain categories are "natural kinds" – real and discovered, not fabricated or invented. Second, people believe that there is some unobservable property, the " 'essence," that causes things to be the way they are. Third, people believe that many everyday words – 'dog,' 'tree,' and 'gold' – map directly onto these natural kinds and therefore have essences. Basic-level category terms are more typically taken to be essentialist in this way than higher or lower level terms (Gelman 2003, 7).

So what does developmental psychology and linguistics tell us about classification in general and biological classification in particular? First, it is inevitable. As we learn a language we automatically learn object categories and hierarchies. Second, it is subjective in that the way we learn language and classify things in the world is dependent on facts about human cognitive and linguistic development. Third, classifications are contextual, depending on the particular culture and environment in which a language is learned. What gets classified and how it gets classified is dependent on which words and phrases are learned and how they are learned. This is surely at least in part a product of how a culture conceives the world – the theories and conceptual frameworks that inform language use. Moreover, functional biases will affect classification based on practical value. Classifications will then vary depending on ways of life.

So does this subjectivity make our classifications arbitrary? Perhaps it does so in some cases, but there is objectivity in our classifications as well. Whole object biases, and shape and function biases, are not *just* products of individual subjectivity, but seem to reflect important facts about the external world. There are, after all, whole objects in the world, and facts about the shapes and functions of these objects – sometimes highly significant facts. Our biases here might have genuine value in navigating the world. If so, then we might ask about an evolutionary explanation. Could our classificatory tendencies and biases have adaptive value and therefore be the products of natural selection? Humans, as well as other creatures, need to distinguish objects, both animate and inanimate, in their environments. *Predator* and *prey*, for instance, are two obvious functional categories that would help in survival. Predators benefit from knowing which kinds of organisms constitute their prey, and conversely, those who serve as prey for others would benefit from knowing what kinds of creatures are potentially predators. If so, then the ability to distinguish different kinds of things could have real value in survival, and not just for humans. Moreover, each sexually reproducing organism needs to distinguish members of its own species from other species if it is to succeed in reproduction. Categorization, in this broad and general sense, is surely of immense value, not just for humans, but for living things in general.

Perhaps there is also a more general epistemic value to classification, as described by Ellen Markman:

> If we responded to each object that we come across as if it were a unique individual, we would be overwhelmed by the complexity of our environment. Categorization, then, is a means of simplifying the environment, of reducing the load on memory, and of helping us to store and retrieve information efficiently. As many investigators have pointed out, categorization is a fundamental cognitive process, involved in one way or another in almost any intellectual endeavor. In identifying objects, in perceiving two things as similar, in recalling information, in solving problems, in learning new information, in acquiring and using language – in all of these cognitive process and more, categorization plays a major role. Forming categories is also one of the major ways in which we learn from experience. Through induction, we note similarities among objects, and are thus able to make important generalizations about the categories. (Markman 1989, 11)

It is not just that classification is useful in identifying particular objects and organisms in our environment; it is useful as a fundamental cognitive process in confronting and making sense of a world of great complexity. It is a way of learning from experience, forming expectations, generalizing, and making inferences.

Suppose these biases are in fact heritable adaptations and provide advantages in survival and reproduction. Are they therefore reliable guides to understanding the world? One possibility is that our categories are useful fictions. They help us navigate the world, and respond to features of the world in useful ways, but don't divide the world in objectively correct ways. They are useful even though they don't really "cut nature at its joints" in the often paraphrased words of Plato. At least some of our categories are subjective in this way. *Weeds*, *pets*, and *vegetables* don't seem to be real categories of things in nature. Instead they seem to reflect preferences relative to our gardens, animal companionship, and nutrition. But another possibility is that many of our classifications do in fact cut nature at its joints, revealing something about the deep structure of nature. One way to approach these two possibilities is as many philosophers have, in terms of a *natural kinds* framework.

Kinds of Kinds

Philosophers typically look to Plato for the origins of the natural kinds framework, most notably in his metaphor in the *Phaedrus* that we should divide the world up as a good butcher divides – by cutting at the joints. The idea is that the butcher who cuts at the joints doesn't just cut wherever he or she wants, but at places determined by the structure of what is being cut. These cuts are easier and neater. We can similarly divide up the world in the places determined by the world – the "joints" of the world, and not by our preferences or desires. When we divide the substances of nature into the elements of the periodic table – hydrogen, oxygen, carbon, and so on, on the basis of atomic number, we are cutting nature at its joints. And when we divide the particles that constitute these elements – protons, neutrons, and electrons, we are also cutting nature at its joints.

Because they result from cutting nature at its joints, *natural kinds* are usually taken by scientific realists to be independent of human beliefs, preferences, conventions, and biases. After all, the structure of the world, and therefore its joints, are presumably independent of human preferences, conventions, and biases. Consequently natural kinds are discovered, not invented. Whether something is an electron or oxygen does not depend on any beliefs, preferences, or desires that anyone may have. By contrast, *nonnatural kinds* are dependent on human preferences, conventions, and biases. They are, in some sense, invented, not discovered. Some nonnatural kinds truly are arbitrary and therefore artificial in the strongest sense, such as *blue foods, cars made on Monday*, or *things bigger than a breadbox*. But some nonnatural kinds seem to be real in that they have become institutionalized or embedded in human practices, laws, and conventions. *Dollar bills, universities, pets, vegetables, farm animals*, and *weeds* are all categories that depend on human conventions, and are in some sense real because of these conventions, but they do not exist in nature independent of the conventions and practices in which they are embedded. A *dollar bill*, for instance, is just a piece of paper in the absence of human conventions. But it is nonetheless a highly significant category for understanding our behavior in the world. And what counts as a *farm animal* is determined by relatively stable farming conventions. There are then, on this way of thinking, three kinds of kinds: *natural kinds* that are independent of human practices and preferences; *conventional kinds* that are dependent on human practices and preferences, but seem real in some sense; and the arbitrary *artificial kinds*.

This framework seems to have the potential to make sense out of the objectivity and subjectivity in the classifications we learn and that we find in various cultures. We can, for instance, ask if the object categories we learn correspond to *natural kinds*, or if they are merely *conventional kinds* or arbitrary *artificial kinds*. How we answer this question has implications about the attitudes we should have toward the various categories. We might plausibly think, for instance, that natural kinds are more important for understanding the world than conventional or artificial kinds precisely because they are real and objective in ways the other kinds are not – by reflecting real divisions in nature. But we might also think that conventional kinds are more appropriate for some sciences. In economics, for instance, conventional kinds, such as *dollar bill, bank*, or *corporation* might be expected to play important roles in a theoretical framework. In these cases, an object category may be subjective in the sense that it depends on human beliefs and conventions, but it is not therefore arbitrary in the way that artificial kinds are.

What about biological categories? Should we think about them as natural, conventional, or artificial kinds? In the middle of the twentieth century biological kinds such as *tigers* and *elms* were often taken by philosophers to be exemplary natural kinds. (See, for instance, Kripke 1972 and Putnam 1973.) But Charles Darwin, in his *Natural Selection*, an unfinished book project that was the basis for his *Origin of Species*, seemed to think of biological species instead as conventional kinds.

> In the following pages I mean by species, those collections of individuals, which have commonly been so designated by naturalists. Everyone loosely understands what is meant when one speaks of the cabbage, Radish & sea-kale as species; or of the Broccoli, & cauliflowers as varieties...
> (Stauffer 1975: 98)

It is certainly possible that many biological categories are conventional, given what we know about, first, biological classifications across cultures, and second, the cognitive and linguistic bases for the learning of object categories. Many of the object terms and categories of cultures are of merely practical or cultural value and need not reflect real divisions in nature. As the ethnobiologists have told us, the Karam of New Guinea don't classify the cassowary as a bird, because of its unique religious significance; the Maya categorize snakes according to whether they are dangerous; and the Tzeltal Indians categorize butterfly larvae, but not adults, according to whether they are edible or dangerous to crops. Many such folk categories are conventional in that they are based on our preferences and practices. And from what we know about the mechanisms for learning object categories in language learning, it is clear that our object categories can be, and often are, merely artificial or conventional. We automatically learn categories and a hierarchy when we learn the words 'red oak,' but we also learn categories and a hierarchy when we learn the words 'red chair.' The latter seems merely conventional in a way the former does not. So are biological categories such as *red oak* and *Homo sapiens* natural kinds, in contrast to the categories such as *red chair* and *dollar bill*? To answer this question we need to say something more about what natural kinds are and how to distinguish them from conventional and artificial kinds.

In recent years there have been extensive discussions about the nature of natural kinds and the use of natural kind terms. We will look at some of the details of these discussions in later chapters, but for now we can start thinking about these kinds of kinds in terms of the standard essentialist approach. On this approach, natural kinds have *essences* that serve as definitions, sets of singly necessary and jointly sufficient conditions that make a thing the kind of thing it is, and that determine what counts as each natural kind. Philosopher of science Elliott Sober explains this approach.

> *Essentialism* is a standard philosophical view about natural kinds. It holds that each natural kind can be defined in terms of properties that are possessed by all and only members of that kind. All gold has atomic number 79, and only gold has that atomic number. It is true, as well, that all gold objects have mass, but *having mass* is not a property unique to gold. A natural kind is to be characterized by a property that is both necessary and sufficient for membership. (Sober 2000, 148)

In this example, the element *gold* has an essence, the atomic number 79, which represents the number of protons found in each atom. Atomic number is determinative in establishing the nature of each substance, and of what counts as that substance. So in this case, the atomic number 79 is necessary and sufficient for gold; something *must have* that number to be gold, and if it does, it is gold. Similarly, the compound *water* has an essence and that is its composition of two hydrogen atoms and one oxygen atom. That structure is necessary and sufficient for something to be water and therefore counts as the essence of, and a definition for water.

Natural kinds, on this essentialist way of thinking, have several important features. First, they are usually understood to be spatiotemporally unrestricted. The essence of gold – its atomic number – is not restricted to any time or place. It applies to substances at anytime and anyplace. So anything that has the atomic number 79 is gold, no matter where and when. That essences are spatiotemporally unrestricted implies, second, that they are unchanging. What makes something gold does not change over time. Now it may be that another substance (perhaps lead) could be changed into gold through manipulation of its atomic structure, but the natural kind itself and its essence has not thereby changed. Third, the essential properties of natural kinds, those that provide definitions and determine what counts as an instance of a natural kind, are usually taken to be *intrinsic*, in that they don't depend on some external state of affairs (Bird and Tobin 2008, 19). The atomic number of gold, for instance, is an intrinsic property of each piece of gold. By contrast, something has *extrinsic properties*, such as being *larger than*, *above*, *descended from*, and *sister of*, only because of some relation to an external thing or state of affairs. Fourth, natural kinds are usually taken to be discrete, in that they do not grade insensibly into each other.

A substance is gold if it has the atomic number of 79, and not 78 or some real number in between. There is therefore a real gap between gold and the other elements.

A Puzzle

If natural kinds are essentialist in these ways, then it is not clear that Linnaean biological taxa, as construed within modern evolutionary theory, can really be natural kinds. First, these taxa are not spatiotemporally unrestricted, or unchanging. *Homo sapiens* appeared at a particular point in evolutionary history and at a particular place. A similar group of organisms that had different origins at a different time and in a different place would be a different species – no matter how humanlike. According to standard practice, this group could not possibly originate in a different lineage and still have the same genus name *Homo*. Moreover, *Homo sapiens* and other biological taxa, change over time. Evolutionary theory tells us that species in general respond to selection forces and change over time without necessarily becoming a new species. The ability to digest lactose and the lactase genes that make it possible, for instance, have spread through certain human populations since the time of the origin of *Homo sapiens*. And certainly the Linnaean tribe Homini and family Hominidai have also changed over time as new member species have appeared and older species have changed or gone extinct.

Moreover, evolution also tells us that species are composed of populations that vary at a particular time. Members of any population of *Homo sapiens*, for instance, vary at a particular time both phenotypically and genotypically. Phenotypic variation is even more striking in *Canis lupus*, the species taxon containing wolves and domesticated dogs. But more importantly it also seems that inclusion in biological categories is not now based on the presence of a set of intrinsic properties at all. An organism is a human if it is born of humans, not if it has a particular set of intrinsic properties. The absence of higher reason, for instance, does not exclude an infant or child from *Homo sapiens*. Similarly, an organism is a dog if it is born of dogs, regardless of its divergence from some normal form. Differences in phenotype and genotype are taken as mere variations among humans and dogs. In actual practice, the essences of biological taxa, if there are essences, are not like the intrinsic properties we see in chemical and atomic kinds. The bottom line is that if members of biological taxa vary over time, and across populations, it is not at all obvious that there is a single set of essential properties that are necessary and sufficient to be that kind of thing.

Moreover, these changes over time, along with variations within population, suggest that many biological taxa may not be discrete. At which point, for instance, did those in the soon-to-be human lineage cease to be an ancestral *hominin* species and become *Homo sapiens*? Anthropologists recognize the difficulty in identifying a boundary, generally giving only a date range for the appearance of modern humans – somewhere between 50,000 and 100,000 years ago. Moreover, if there has been substantial interbreeding between *Homo sapiens* and *Homo neanderthalensis*, as many think, are they still two *discrete* species? (Or should we regard them both as subspecies of *Homo sapiens*?) The offspring of some romantic liaison would presumably be a member of two different biological taxa. If so, then it isn't clear that these taxa are fully discrete. As we will see in Chapters 5 and 6, there is substantial hybridization between plant species and massive horizontal gene transfer among bacteria and viruses. In each of these cases it is not clear that the biological taxa are discrete in the way chemical and atomic kinds are discrete.

This all suggests that biological taxa cannot be *natural kinds* – at least on the standard essentialist account of natural kinds. But if they aren't natural kinds, are they then conventional kinds, as Darwin seemed to suggest in the passage quoted earlier? If so, then biological classification would be merely conventional at best, or arbitrary at worst.

But the categories in modern biological classification do not seem to be arbitrary or conventional in this way. *Homo sapiens* really is distinct in important and objective ways from other biological categories such as *Canis lupus*. The distinction between humans and dogs and wolves does not seem to be a mere convention. Nor does it seem to be arbitrary.

The puzzle is this. Modern biological classification seems to divide the world up into taxa that have some features of natural kinds. They seem to have objective reality in that they don't depend on subjective preferences or practices. They are not arbitrary groupings but somehow seem to carve nature at its joints. In many ways the biological categories containing humans, tigers, dogs, birds, and bacteria seem a lot like the natural kinds oxygen, gold, water, protons, and electrons. But if they cannot be natural kinds, how can we nonetheless understand the apparent objectivity of biological classification? This is a question that will be lurking through the remaining chapters of this book, and we will return to it in more detail in Chapter 7. But to fully understand the modern scientific practice of biological classification, we need to look at the history of that practice – how it began and evolved into its modern form. That is the topic of the next three chapters.

2 The Aristotelian Framework

The Essentialism Story

According to one standard history of biological classification, prior to Darwin biological taxa were conceived as the timeless and unchanging essentialist natural kinds introduced at the end of Chapter 1. The original basis for this essentialism, this history tells us, was Aristotle's method of logical division that was accepted and developed by the naturalists who came later. The philosopher Daniel Dennett is one prominent advocate of this interpretation of the history of classification:

> Aristotle had taught, and this was one bit of philosophy that had permeated the thinking of just about everybody, from cardinals to chemists to costermongers ... [that] All things – not just living things – had two kinds of properties: essential properties, without which they wouldn't be the particular kind of thing they were, and accidental properties, which were free to vary within the kind ... With each kind went an essence. Essences were definitive, and as such they were timeless, unchanging, and all or nothing. (Dennett 1995, 36)

According to Dennett, this form of essentialism ruled until the time of Darwin:

> The taxonomy of living things that Darwin inherited was thus a direct descendant, via Aristotle, of Plato's essences ... We post-Darwinians are so used to thinking in historical terms about the development of life forms that it takes a special effort to remind ourselves that in Darwin's day species of organisms were deemed to be as timeless as the perfect triangles and circles of Euclidean geometry. (Dennett 1995, 36)

Alongside Dennett here are perhaps the majority of biologists and philosophers who concern themselves with the history of classification (Stamos

2007, 137). Marc Ereshefsky, for instance, associates essentialism with virtually all of the major naturalists who preceded Darwin:

> [T]he common philosophical stance towards species was essentialism. Linnaeus, John Ray, Maupertuis, Bonnet, Lamarck, and Lyell all adopted an essentialist (or typological) view toward systematics ... On this view, classification systems highlight natural kinds – groups whose members share kind-specific essences ... According to biologists who adopted an essentialist view of species, species have the same role in biological taxonomy as the chemical elements have on the periodic table. All and only the members of a species taxon should have a common essential property. (Ereshefsky 1992, 188–189)

As the end of Chapter 1 makes clear, this history, if it were accurate, would at least make sense of how pre-Darwinians might think their biological classifications to be natural and objective, by "cutting nature at its joints." Biological taxa would be essentialist natural kinds.

There are, as we shall see, good reasons to be skeptical of this history, dubbed by historian of science Mary Winsor "the essentialism story." She argues that it was fabricated by the biologist Ernst Mayr, with the help of another biologist, Arthur Cain, and a philosopher, David Hull:

> The essentialism story is a version of the history of biological classification that was fabricated between 1953 and 1968 by Ernst Mayr, who combined contributions from Arthur Cain and David Hull with his own grudge against Plato. It portrays pre-Darwinian taxonomists as caught in the grip of an ancient philosophy called essentialism, from which they were not released until Charles Darwin's 1859 *Origin of Species*. Mayr's motive was to promote the Modern Synthesis in opposition to the typology of idealist morphologists; demonizing Plato could serve this end. Arthur Cain's picture of Linnaeus as a follower of 'Aristotelian' (scholastic) logic was woven into the story, along with David Hull's application of Karl Popper's term, 'essentialism,' which Mayr accepted in 1968 as a synonym for what he had called 'typological thinking'. (Winsor 2006, 149)

Mayr was anxious to contrast the "modern synthesis" of population genetics and Darwinian natural selection, which treats variation within populations as an important part of the evolutionary process, with the essentialism that seemed to deny or dismiss this variation. The portrayal of Aristotle, Linnaeus, and others as essentialists who treated variation as unimportant or illusory helped distinguish this new population thinking from the view it was supposed to replace. In effect, the essentialism story made the modern synthesis seem even more revolutionary, replacing a 2,000-year-old outdated essentialist worldview.

The essentialism story is misleading at best. The history of biological classification, beginning with Aristotle and continuing through to Darwin, is not a simple history of essentialist thinking about biological classification. It isn't clear, for instance, that Aristotle or Linnaeus, the two arch essentialists in the essentialism story, were essentialists about biological taxa in the assumed way at all. Nonetheless, this essentialism story has been widely accepted, even in the face of contrary evidence from primary sources and the skepticism of various historians and philosophers (Winsor 2003, 389–390). To understand the development of modern biological classification, we need to set aside this essentialism story and look more carefully at the views of Aristotle and those who followed. We shall see that the story of pre-Darwinian classification is at minimum much more complicated than the essentialism story would have us believe.

The Aristotelian Framework

If classification is generated automatically in the use of general terms, then the history of biological classification should perhaps start with the development of language. But that is a history that cannot be written with any confidence. We simply lack evidence about the early development and use of language. We can begin that history with more confidence, though, in the thinking of the Ancient Greeks, Plato and Aristotle in particular. Aristotle (384–322 BC) is certainly the primary figure in the early history of biological classification, but it is worthwhile to briefly look at Plato, his mentor and foil. There are three doctrines advocated by Plato that played a role in the thinking of Aristotle and those who followed. First is the famous proclamation from his *Phaedrus*, referenced in Chapter 1, that we should follow the example of a good butcher and "cut nature at its joints" (Plato 1961, Phaedrus 265d–266a). This is usually understood to demand at minimum that when we divide and group things in nature through classification, we should divide at the "natural" divisions, producing the groupings and divisions determined by nature, rather than our own interests. The second doctrine is the method of dichotomous division or "diairesis" that was to do this "cutting at the joints." Plato illustrated this method of division with angling. First, he divided the *arts* into two categories, the *productive arts* and the *nonproductive (acquisitive) arts*, then divided the latter category into *acquisition by exchange* and *acquisition by coercion. Acquisition by coercion* was then divided into *open acquisition (fighting)* and *secret acquisition (hunting)*. This division continued until after a long discourse we find that angling is a subcategory of *hunting of live things, hunting of animals in water (fishing), fishing by blows, day fishing*, and finally *with blows from below* (Plato, Sophist 219c–221d; see also Wilkins 2009,14–15).

The third Platonic doctrine was the theory of forms. According to Plato, the things in this world that we see and experience are just imperfect copies of things in an ideal, transcendent, intelligible world. Plato called these things in the intelligible world 'eide,' (singular 'eidos'), typically translated into English as 'forms' or 'ideas.' Plato claimed that we are acquainted with things in the world around us through sensation, but come to have true knowledge only through reason. We can know an individual physical horse, for example, through our senses - what it looks like, sounds like, feels like, and so. But for true understanding, we need to understand the form *horse* through reason. In part this is an understanding of the body plan of horses in general. (What else this is supposed to involve is the subject of much Plato scholarship. See Mason 2010, 27–59.) We can then understand the general term 'horse' based on our understanding of the form of *horse*. If so, when we learn the term 'horse' we are really learning something about the form underlying all individual physical horses. This theory of forms is the metaphysical basis for the idea that we can cut nature at its joints. When we cut or divide nature so that we have cut between these forms we have divided correctly. Grouping all physical instantiations of the form of horse together into the category horse, and not grouping other things with them, zebras and donkeys for instance, would perhaps be to cut nature at its joints. While Plato thought that there were forms for the various kinds of living things, he also applied this doctrine to nonliving things and artifacts such as tables and chairs, and abstract things such as beauty, justice, and the good (Mason 2010, 40–43).

Aristotle was Plato's student, but like a good student he disagreed with many of his mentor's views. In particular he disagreed with Plato's claim that the forms (*eide*) exist separately from physical things in an intelligible realm. But Aristotle did adopt and develop some aspects of Plato's framework in his method of logical division, which is assumed by the essentialism story to serve as the foundation for the alleged pre-Darwinian essentialism in biological classification. (See, for instance, Ereshefsky 2001, 201–202.) The basis for this method is found in Aristotle's use of the Greek terms 'eidos' and 'genos' (translated into the Latin as 'species' and 'genus'). The idea is that to determine the essence of a thing, it is necessary to determine first its *generic* nature and second, its *specific* attributes – the *differentia* that distinguish it from other kinds of things that share the generic nature. The definition of a species (*eidos*) would then be the set of properties that are necessary to be a member of its genus, and a property or set of properties that differentiate it from the other species in that genus. According to the essentialism story, this method was used to determine the essences of species. Typically this method is also taken to be hierarchical and dichotomous, whereby the differentia are understood as the presence or absence of a trait. Marc Ereshefsky explains:

> According to the method of dichotomous division, a proper classification consists of a hierarchy of classes, each defined by the genus it belongs to, and its *differentia* within that genus. Furthermore, a genus should be dichotomously dividing according to which entities have a particular *differentia* and which do not. The genus of animate objects is divided into animals and vegetables according to the *differentia* of self-movement. Moving down one level in the hierarchy, the genus animal is divided into man and the lower animals according to the *differentia* of rationality. (Ereshefsky 2001, 20)

According to this picture of the logic of division, things can be classified into a multilevel hierarchy of genera and species based on the presence or absence of general traits and more specific differentia.

This interpretation of Aristotle is accurate insofar as he applied this method in his logical works. In his *Posterior Analytics*, for instance, he applied the method of logical division to a variety of subjects, including numbers and geometric figures.

> When you are dealing with some whole, you should divide the genus into what is atomic in species – the primitives – (e.g. number into triplet and pair); then in this way attempt to get definitions of these (e.g. of straight line and circle and right angle); and after that, grasping what the genus is (e.g. whether it is a quantity or a quality). (Aristotle 1995, 96b20)

There he also claimed that this method could be used at multiple levels.

> We should look at what are similar and undifferentiated, and seek, first, what they all have that is the same; next, we should do this again for other things which are of the same genus as the first set and of the same species as one another but of a different species from those. And when we have grasped what all these have that is the same, and similarly for the other, then we must again inquire if what we have grasped have anything that is the same - until we come to a single account. (Aristotle 1995, 97b7–13)

At some particular level of analysis that which is similar and undifferentiated relative to some attribute (property) or set of attributes will constitute a genus. And those things differentiated from others within the genus, but similar among themselves, will be a species. Then that species can in turn be treated as a genus and we can look for some subset within it that is again differentiated into lower level species.

But according to the essentialism story, and as we have just seen in the examples given by Ereshefsky, Aristotle's method of logical division was applied not just to things like numbers and geometric figures, but also to living creatures, grouping them into *eide* (species) and *gene* (genera), based on the presence or absence of such traits as self-movement and rationality. There are several problems with this story, however. The first problem is that in his *Parts of Animals*, Aristotle explicitly denied that the demarcation of animal kinds is in fact based on dichotomy.

> One should try to take animals by kinds, following the lead of the many in demarcating a bird kind from a fish kind. Each of these has been defined by many differences, not according to dichotomy. (Aristotle 1995, e643b9–12)

Aristotle rejected dichotomous division because, first, many of the criteria we might use are superfluous or unimportant. The differentiae *cleft-footed*, *two-footed*, and *footed* might all be used, for instance, but only the first was regarded by Aristotle as significant (Aristotle 1995, 642b8–10). Second, privative terms, based on the mere absence or presence of a trait, cannot be subdivided further. One cannot divide *footless* into further subdivision of footlessness. Privative traits therefore lead to dead ends in division.

> [P]rivative terms inevitably form one branch of dichotomous division, as we see in the proposed dichotomies. But privative terms admit of no

> subdivision. For there can be no specific forms of negation, of Featherless for instance or of Footless, as there are of Feathered or Footed. Yet a generic differentia must be subdivisible; for otherwise what is there that makes it generic rather than specific? There are to be found generic, that is specifically subdivisible differentiae; Feathered for instance and Footed. For feathers are divisible into Barbed and Unbarbed, and feet into Manycleft, and Twocleft. (Aristotle 1995, 642b22–30)

Third, many dichotomous attributes divide into unnatural groups (emphasis added):

> It is not permissible to break up a natural group. Birds, for instance, by putting its members under different bifurcations, as is done in the published dichotomies, where some birds are ranked with animals of the water, and other placed in a different class ... If such natural groups are not to be broken up, *the method of dichotomy cannot be employed, for it necessarily involves such breaking up and dislocation.* (Aristotle 1995, 642b10–020)

Although it is not precisely clear in this passage what makes some classification "natural" for Aristotle, it isn't too hard to see how the subcategory (*eidos*) *waterbirds*, based on the attribute of *living in water*, placed in a more inclusive category (genus) *animals of the water* would be problematic. *Waterbirds* would then be categorized with *fish, sea otters*, and *crabs*, and not with *forest birds*. Aristotle concluded his criticism here with the outright rejection of dichotomy: "It is impossible then to reach any of the ultimate animal forms by dichotomous division" (Aristotle 1995, 644a12–13).

If we cannot use dichotomous division to classify organisms, how can we do so? Aristotle doesn't say with any specificity, but suggests that general similarity forms larger groups, and differences in degree of attributes serve to generate the subdivisions.

> It is generally similarity in the shape of particular organs, or of the whole body, that has determined the formation of larger groups. It is in virtue of such a similarity that Birds, Fishes, Cephalopoda and Testacea have been made to form each a separate class. For within the limits of each such class, the parts do not differ in that they have no nearer resemblance than that of analogy – such as exists between the bone of a man and the spine of a fish – but differ merely in respect of such corporeal conditions as largeness smallness, softness hardness, smoothness roughness, and other similar oppositions, or in one word, in respect of degree. (Aristotle 1995, 644b8–15)

Birds, for instance, all seem to share some general similarity in terms of many attributes: their bipedal posture, beaks, feathers, and wings. Among those things with this general similarity are subdivisions based on differences in degree in leg length, beak size and shape, wingspan, and so on. The *genos* grouping here would not be based on a single attribute, and the subdivisions into *eide* would not be based on a single differentia, but on many differences of degree. This is clearly not the application of the method of logical division assumed by the essentialism story.

But perhaps an even more significant problem for the method of logical division is that many creatures have equivocal attributes: they cross-divide based on resemblances to different classes, such as animals and vegetables. Aristotle recognized this fact (emphasis added):

> In the sea, **there are certain objects concerning which one would be at a loss to determine whether they be animal or vegetable.** For instance, certain of these objects are fairly rooted, and in several cases perish if detached; thus the pinna is rooted to a particular spot, and the razor-shell cannot survive withdrawal from its burro. Indeed, broadly speaking, the entire genus of testaceans has a resemblance to vegetables. (Aristotle 1995, 588b13–17)

The organisms in *Pinna* (a type of bivalve mollusc), for instance, share certain attributes with animals - being fleshy and having digestive tracts, but they also share attributes with plants - they are rooted and often die if uprooted. Using the logic of division, the members of *Pinna* could be either animal or plant depending on which attributes are selected - those shared with animals or those shared with plants. This method is therefore equivocal. The historian David Balme seems to have been right when he concluded that Aristotle did not think that division was capable of generating a satisfactory hierarchical classification in living things (Balme 1987, 84).

There are additional reasons to doubt the essentialism story about Aristotle's method of logical division. First, because this method was used to classify nonbiological things - numbers, mathematics, and so forth - it could not be a *biological* system like the modern Linnaean system. It was an all-purpose, *logical* system of classification that was not tied explicitly to any natural or nonnatural domain. Second, because the terms 'eidos' and 'genos' were used at multiple levels, depending on the investigation at hand, there were no fixed hierarchical levels. For instance, in a single passage from the *History of Animals*, Aristotle treated as *gene* the blooded animals, the quadrupeds that give live birth, and those that lay eggs, as well as fish, birds, and cetaceans (Aristotle 1995, 505b25–35). But elsewhere he treated quadrupeds and birds as *eide* (Aristotle 1995, 99b3–6). *Genos* (genus) and *eidos* (species) are clearly not limited to fixed, single levels, and certainly not the levels that correspond to modern taxonomic usage (See also Lennox 1980, 324, fn. 11). The only constant seems to be a relational one. In *genos–eidos* (genus-species) comparisons, the *genos* indicates a more inclusive level of grouping than species. The *eidos* of Aristotle is certainly not just the *species* of modern Linnaean classification.

The third problem is that Aristotle's main project in his biological works didn't seem to be classificatory, but functional and explanatory. Instead of observing the essential attributes of animals to group them into animal kinds, he was more interested in discovering the general principles of the distribution of attributes or *parts* among organisms – and in order to give explanations.

> The course of exposition must be first to state the essential attributes common to whole groups of animals, and then to attempt to give their explanation. Many groups, as already noticed, present common attributes, that is to say, in some cases, absolutely identical–feet, feathers, scales, and the like; while in other groups the affections and organs are analogous. For instance, some groups have lungs, others have no lung, but an organ analogous to a lung in its place; some have blood, others have no blood, but a fluid analogous to blood, and with the same office. To treat of the common attributes separately in connexion with each individual group would involve, as already suggested, useless iteration. For many groups have common attributes. (Aristotle 1995, 645b1–14)

In this passage, when Aristotle used the term 'group' he was referring not to what we would identify as species or genera, but instead to groups of animals with particular functional parts such as hearts and lungs. What is important is not *that* particular kinds of organisms have a particular attribute such as a heart, but *how* that attribute gets correlated with other kinds of attributes such as lungs in *whatever kinds of organisms* they are found.

This is an explanatory project in that when Aristotle discussed necessary conditions or essences, he was not referring to the logical requirements in a definition of a kind of animal or plant, but functional requirements for organisms to survive and flourish. James Lennox highlighted this fact:

> By far the central question on Aristotle's mind throughout the discussion is *to what end* do these variations exist. Why is the hawk's beak hooked and why are its talons sharp and curve? Why are a duck's toes united by webbing, and why is its beak wide and flat? For what purpose are the legs and toes of cranes long and thin, and is there some relationship between these facts and cranes' relative dearth of tail feathers and wings? (Lennox 1980, 341)

What is important are the functional relations among the parts of an organism, an environment, and a way of life.

> [A]t least in regard to living things, the 'essence/accident' distinction is a distinction between those features which are required by the kind of life an animal lives and those which aren't. If a crane is to survive and flourish, it *must* have, not simply 'long' legs, but legs of a certain length, defined relative to its body, neck length, environment, feeding habits, and so on. (Lennox 1987, 356)

The bottom line is that the essential properties here are not *definitionally* necessary, but are *functionally* necessary for a particular lifestyle in a particular environment. Aristotle was not so much interested in classifying living things based on their parts as he was in understanding how the parts of living things function together.

Finally, to fully understand Aristotle's views we need to recognize that he used the terms 'eidos' and 'genos' in multiple ways. According to Philip Sloan there were three different uses of these two terms. The first was as logical universal.

> Understood as universals ... Aristotle's concept of species and genera carry no particular biological significance. Any entity is subject to the predication of its *gēnos* and *ēidos*, and there is no particular restriction of the usages to living beings. (Sloan 1985, 103)

This use is primarily in Aristotle's logical works, and is the sense discussed in the context of the method of logical division. Here the *eidos* was simply a subdivision of a *genos* – at some level or other in an inquiry, and forms a hierarchy whereby the *eidos* at one level can be a *genos* at another. It is in this usage that the essentialism story is most plausible, even though, as we have just seen, Aristotle denied that logical division could get us the animal kinds.

On another sense, which we find in Aristotle's biological works, the *eidos* was an immanent or enmattered form:

> As enmattered form, *ēidos* is not *per se* a universal or logical class, but constitutes the primary being of an empirically discernible entity. As form, *ēidos* is both individual and non-material. In its pure individuality – as the immanent principle of shape, structure, intelligibility and order of a particular individual thing – it is not expressible in language, nor definable in itself. (Sloan 1985, 104)

This sense of the *eidos/genos* pair is not hierarchical and is not worked out in terms of the method of logical division. Instead the *eidos* is the material form or shape of a thing, perpetuated in generation. Here the term 'genos' can be understood in terms of origins (as in 'genesis') and relative to living things, such as a lineage, race, or family. An *eidos* is a shape, form, or appearance of a thing that has its origins in a lineage, race, or family (Peters 1967, 46).

We can think of this in terms of family resemblance. Children tend to have a particular form or appearance similar to, and derived in generation from their parents. We can see this appearance perpetuated over time within this family lineage. This use of these terms could also be understood in the modern biological sense, applying to a bird species, whereby there is a form, shape, or resemblance among all birds of that particular species that has been perpetuated in reproduction. Philip Sloan claimed this is the primary meaning we find in Aristotle's biological works (Sloan 1985, 104). If so, then the essentialism story is misleading when it looks only to the logical sense of 'eidos' and 'genos' in understanding Aristotle's use of these terms in his discussion of living things.

The third use of 'eidos' and 'genos' may have originated with Aristotle (Peters 1967, 50). Here the *eidos* was the principle of organization or development and the *genos* the matter or system of shared characteristics out of which species get differentiated. Among birds, for instance, there is a shared set of traits out of which the different specific forms develop. All members of Aves have beaks, legs, and feathers and a similar general appearance, but in the different *eide* (species) the beaks develop to a different length and shape, as do the legs and feathers. These differences are just a matter of degree – more or less – rather than discrete. And the functional value is critical, as Lennox explained.

> Viewed in abstraction from their specific environments, the variations between the beaks, wings, feathers, or legs of different birds appear to be nothing but variations in degree of the common differentiae of the genus of birds. But when viewed as adaptations to a peculiar mode of life, it becomes clear that each variation is the proper and peculiar one for a given species' way of life; no other differentiation would serve the needs of the species. (Lennox 1980, 342)

On this way of thinking, the *eidos* is a specific development of a more general body plan, the *genos*. Here the essential traits are based on the functional role they play in a lifestyle – based on their variation in form. The beaks of the Galapagos finches, which so influenced Darwin's thinking, provide a good example of this sense of *eidos* and *genos*. There is a general body plan among the finches (a *genos*) that forms the basis for the development of different kinds of beaks in terms of form – size, length, and shape (specific *eide*). Here the specific size, shape, and length of a beak are essential in that they are necessary for a certain way of life for the finches, based on what kinds of seeds are available – larger and harder seeds versus smaller seeds. In Aristotelian terms, what each form eats and how it lives is its proper activity. And the *genos* develops into specific *eide* to serve this proper activity, as Lennox indicated:

> The "form" of living things is their proper activity, and the generic material is differentiated into a species member for the sake of this. What ensures the integrity and reality of Aristotelian species – in spite of the fact that their organs differ structurally only in degree along a variety of continua – is that the members of each species are individuals adapted to a specific manner of life. (Lennox 1980, 343)

Lennox then concluded: "with its stress on the way in which the requirements of adaptation determine the precise differences between species within a genus, it is a spiritual ancestor of the evolutionary approach" (Lennox 1980, 345). Notably, and whether or not Aristotle's approach is the spiritual ancestor of an evolutionary approach, when Darwin finally read Aristotle near the end of his life, he was surprised and impressed by what he found. In an 1882 letter written to William Ogle, who had sent him a copy of his translation of Aristotle's *Parts of Animals*, Darwin wrote:

> From quotations which I had seen I had a high notion of Aristotle's merits, but I had not the most remote notion of what a wonderful man he was. Linnaeus and Cuvier have been my two Gods, though in very different ways, but they were mere school-boys to Aristotle. I never realized before reading your book to what an enormous summation of labor we owe even our common knowledge. (F. Darwin 1887, 427)

So what does all of this imply about Aristotle and his role in biological classification? First, his method of logical division *could* be used to construct an essentialist system of biological classification. But Aristotle did not in fact use this method in his biological works, and gave reasons why it wouldn't generate a satisfactory classification. The characterization of him as an essentialist relative to *biological* classification is therefore misleading. Second, Aristotle's use of the terms 'eidos' and 'genos' to refer to enmattered forms perpetuated in generation suggests that he was also thinking of organisms as parts of lineages. As we shall see, this idea has much in common with the modern way of thinking about biological taxa. *Homo sapiens*, for instance, has a particular "enmattered form" that has been perpetuated in a lineage for the last 50,000–100,000 years. And there is an even more general and less specific form that humans share with other primates, and that seems to have been perpetuated by generation in the primate lineage for a much longer time. Finally, there is the third sense of 'eidos' as a specific developmental tendency for differentiation out of a more general 'genos.' Here specific forms are developmental adaptations to lifestyles and environments. The long legs of wading birds, for instance, are a specific development to a particular lifestyle and environment, from a more generic bird body plan. These nonlogical senses of 'eidos' and 'genos' are at least part of the reason Lennox claimed Aristotle's way of thinking as a "spiritual ancestor of the evolutionary approach." Whether or not Lennox was right about this, the portrait of Aristotle painted by the essentialism story is clearly misleading. He was not a simple essentialist using the method of logical division to define species and genera with timeless and unchanging essences.

What is also important here is that the application of the predicates Aristotle used in thinking about the parts of living things – as being two-footed, or having hearts and lungs, spontaneously generates classifications by treating them as more general *gene* or more specific *eide*, as they become categories of things that are two-footed and have hearts and lungs. These categories were important for Aristotle in understanding how certain functional attributes such as having a heart and having lungs are coextensive. This is part of understanding the functional integration of organisms in general and across biodiversity. So while *eidos* as enmattered form may be important for one inquiry, *eidos* as a subdivision of some *genos* may also be important relative to another. For Aristotle there was, however, no *single* overarching, hierarchical classification such as we see in the modern evolution-based Linnaean classification. And there would be no obvious theoretical reason for Aristotle to postulate such a single classificatory scheme. This is in contrast to modern biology, which, as we shall see, is motivated by the idea that all living things have common origins in a single evolutionary tree.

The Transformation of Aristotle

Even though there are good reasons to doubt that Aristotle was the arch-essentialist portrayed by the essentialism story, there is a long history of commentary that has interpreted him in ways that are friendly to this portrayal. In part this is due to the fact that those who followed were less interested than Aristotle in the empirical study of nature in general, and living things in particular. One exception was his student Theophrastus (371–287 BC), who engaged in extensive studies of plants and is taken by some to be using Aristotle's method of division to classify plants (Wilkins 2009, 21–24). Beginning in the first century AD, though, knowledge of Aristotle's work was primarily through a series of commentators, who had other agendas, in particular reconciling Aristotle with Plato or with Christian theology. But most significant for the purposes here, these commentators had little or no familiarity with, or interest in Aristotle's biological works, focusing mostly on his *Categories* and *On Interpretation*, works that were focused on language. Aristotle's treatment of *eidos* and *genos* as enmattered forms perpetuated in generation and as developmental principles was simply unknown to these commentators. Aristotle was transformed from a biologically informed, comprehensive thinker into a logician and linguistic philosopher.

Little is known about the fate of Aristotle's works after his death in 322 BC. One story is that he gave them to his student Theophrastus, who in turn passed them on to a nephew, Neleus, who went to Scepsis and hid them in a cave. These manuscripts were then discovered two centuries later, taken first to Athens and then to Rome, where Andronicus organized them according to topic. They were the likely source for modern compilations as well as material for the commentators who followed (Barnes 1995, 11).

The first, "authentic" commentators were mostly interested in getting Aristotle's views right, but also in developing and extending those views. In the first century AD, the most important of these early commentators, Alexander of Aphrodisias, wrote on many of Aristotle's works, but not the biological. He must have had some familiarity with the biological works though, referring to them in other commentaries (Madigan 1994, 81). Like other early commentators he treated Aristotle's work as a unified and consistent whole. But he also recognized the different senses of 'eidos' and 'genos' in Aristotle's work, not just as logical universals, but as enmattered forms and developmental principles (Madigan 1994, 87). Recognition of these different uses gets lost in subsequent commentaries.

The Neo-Platonist commentators who followed were less interested in understanding Aristotle than in reconciling him with Plato. Plotinus (204/5–270 BC), one of the first in this tradition, was interested in Aristotle primarily because of his direct knowledge of Plato's views, but Plotinus didn't assume that Aristotle and Plato were in agreement. Porphyry (234–305 AD), who edited Plotinus's works and wrote *Life of Plotinus*, was most responsible for how the later tradition of commentary developed (Ebbesen 1990b, 141; Sorabji 1990, 17). Porphyry claimed that Plato and Aristotle were in agreement, as indicated by the title of one his works: "On the School of Plato and Aristotle being One" (Sorabji 1990, 2). Aristotle then became the starting point for discussions and commentaries on Plato. Porphyry's interpretation of Aristotle's method of difference was later used to construct the dichotomous "tree of Porphyry." Here *substance* was divided into corporeal and incorporeal. Corporeal substance was then divided into inanimate and animate, which was then divided into insensitive and sensitive. The latter was divided into irrational and rational, which was then divided into Socrates, Plato, and Aristotle (Wilkins 2009, 29–32). Porphyry also established Aristotle's *Categories* as the standard starting point for the study of Aristotle (Blumenthal 1996, 22; Falcon 2005, 8).

The title of Aristotle's *Categories* comes from the Greek term for predicate, 'kategoreisthai.' Superficially it seems to be about just that - language and the application of predicates to things in the world. In the first section, the *Pre-Predicamenta* (after the Latin word for predicate), Aristotle distinguished primary and secondary substances. Primary substances are the individual things to which the predicates apply, but are not themselves predicates: "A substance - that which is called a substance most strictly, primarily, and most of all - is that which is neither said of a subject nor in a subject, e.g. the individual man or the individual horse (Aristotle 2a13–15). Secondary substances are those that are predicated of primary substances. An individual human, Socrates for instance, would be a primary substance. He was not a predicate but a thing to which predicates apply. Socrates was 'animal', 'bipedal', 'human', 'rational',' social,' and 'political.' Similarly, a particular bird is a primary substance to which a series of predicates might apply: 'animal,' 'wading bird,' 'long-legged,' 'eats fish,' and so on.

Aristotle then gave a division based on whether something is "being said of" or "being in." The primary substances, individual things, are said not to be in a subject or of a subject. Secondary substances, in contrast, are not in the subject, but can be said of the subject. 'Man' and 'animal,' for instance:

> For man is said of the individual man as subject but is not in a subject: man is not in the individual man. Similarly, animal also is said of the individual man as subject, but animal is not in the individual man. (Aristotle 1995, 3a 8–15)

The differentia, like secondary substances, are also not in the subject, but are said of it: "For footed and two-footed are said of man as subject but are not in a subject; neither two-footed nor footed is in man" (Aristotle 1995, 3a 23–25).

What is important here is that we come to know the primary substances - individual things - through secondary substances or predicates.

> It is reasonable that, after the primary substances, their species and genera should be the only other things called secondary substances. For only they, of things predicated, reveal the primary substance. For if one is to say of the individual man what he is, it will be in place to give the species or the genus (though more informative to give man than animal); but to give any of the other things would be out of place - for example to say white or

> runs or anything like that. So it is reasonable that these should be the only other things called substances. (Aristotle 1995, 2b29–36)

This reveals the significance of Aristotle's distinct uses of 'eidos' and 'genos.' We can know an individual object (primary substance) as an *eidos* in the enmattered form sense, through mere sensation – its appearance or what it looks like. But for higher knowledge of a thing, we need to know it as *eide* and *gene* in the logical universal sense. We need to know how it is similar to and different from other kinds of things in more general categories. We need to know the patterns of predicates that apply to it. Notice that in the *Categories* the predicates – the eide/species and gene/genera – spontaneously generate classifications. As Aristotle understood, our use of universal terms divides and groups things in the world, based on how we apply these terms. To say that something is 'animal', 'human,' or 'rational' is to categorize it as an animal thing, a human thing, or a rational thing.

As of yet there is still no consensus about how to interpret *The Categories*, whether it is just about language and patterns of predication, about things and the nature of reality, or about the concepts that ground linguistic practices. In part this is because Aristotle often repeatedly shifts locution from "what is" to "what is said." This ambiguity in *The Categories* would play out in the debate about universals of the next millennium.

Porphyry wrote several commentaries on *The Categories*. In them he made Aristotle compatible with Plato by interpreting *The Categories* as being just about language and predication, and limited to just the material world. He therefore avoided the possibility that Aristotle's views could conflict with Plato's theory of transcendent Forms (Ebbesen 1990b, 145–146). In his *Isagoge*, or "introduction" to the *Categories*, he also set up what would become for later discussions the "problem of universals."

> At present ... I shall refuse to say concerning genera and species whether they subsist or whether they are placed in the naked understanding alone or whether subsisting they are corporeal or incorporeal, and whether they are separated from sensibles or placed in sensibles and in accord with them. Questions of this sort are most exalted business and require great diligence of inquiry. (Jones 1969, 186)

Here he gave us three sets of possibilities: first, whether universals (species and genera) subsist – exist outside of the mind; second, if they exist outside of the mind, whether they exist as physical or nonphysical things; third, whether they exist separately or only in perceptible things. This formulation of the problem was adopted by the Christian commentator Boethius.

Boethius, born near the end of the fifth century AD, translated some of Aristotle's logical works and wrote commentaries on his *Categories* and *On Interpretation*, as well as on Porphyry's *Isagoge* (Ebbesen 1990a, 374). These works, which relied heavily on Porphyry's analysis, came to be the standard source for Aristotle scholars through the twelfth century, when Aristotle's texts were "rediscovered" through Arabic sources (Marenbon 2005, 15). Boethius responded to Porphyry's statement of the problem of universals by arguing that universals subsist in sensible things, but are also understood apart from bodies, and that they are not mere constructions of mind, but represent reality (Marenbon 2005, 3–4). This is consistent with the Platonic view that there is a timeless reality independent of enmattered things (Hyman and Walsh 1983, 115). According to Boethius, there was an isomorphism between the universal terms – the *species* and *genera*, as concepts and things in the world.

The Neo-Platonism that we see in Boethius had already been combined with Christian theology in the thinking of Augustine (354–430 AD), who interpreted Plato's metaphysics within a Christian framework, where *species* were ideas in the mind of God. *Species* then take on an explanatory role, as "reasons":

> In Latin we can call the Ideas "forms" or "species," in order to appear to translate word for word. But if we call them "reasons," we depart to be sure from a proper translation – for reasons are called "logoi" in Greek, not Ideas – but nevertheless, whoever wants to use this word will not be in conflict with the fact. For Ideas are certain principal, stable, and immutable forms or reasons of things. They are not themselves formed, hence they are eternal and always stand in the same relations, and they are contained in the divine understanding. (Klima 2003, 197)

The idea seems to be that *species* are the ideas in the mind of God at creation, and can therefore explain how nature came to be as it is. In this sense, *species* are the principles that governed creation and serve as exemplars for all the enmattered forms we see in the world. This grounds the Neo-Platonist isomorphism between words, concepts, and things. The ideas of God determine the things in the world to which we apply our terms, and that serve as the basis for concepts in the understanding. In other words, for Augustine the classifications of enmattered things in the world, the *species* and *genera*, represent God's ideas that governed creation. To know the names of things, the *species* and *genera*, is therefore ultimately to know the mind of God.

Universals

After Boethius there was a long philosophical debate about the nature of universals. This debate usually occurred within the framework outlined by Porphyry, who had asked, first, whether species and genera subsist or are in the understanding alone; second, if they subsist whether they are corporeal or incorporeal; third, whether they are separate from or in sensibles. Three main kinds of answers to these questions have typically been distinguished: realism, nominalism, and conceptualism. (This is surely an oversimplification. See Thompson 412, 418–419.)

The Christian Neo-Platonists, like the non-Christian Neo-Platonists, tended to be realists. But instead of believing that universals exist as *forms* in a transcendent realm, the Christian Neo-Platonists believed that universals exist as timeless ideas in the mind of God. On both Neo-Platonist views, though, universals are independent of the particulars in which they are instantiated. Both the forms and the ideas in the mind of God exist independently of things in the world. This was a position held by many theologians, including John Scotus Erigena in the ninth century, William of Champeaux and St. Anselm of the eleventh and twelfth centuries, and Henry of Ghent in the thirteenth century (Jones 1969, 187; Thompson 1995, 409–414; Klima 2003, 199). This realism had some significant theological virtues. If universals were real, then the term 'God' could be a universal, referring to something real and be instantiated in the three persons of the Trinity, and yet still be "one." And if the term 'man' referred to something real and independent of its instantiation in particular men, then something that applied to the universal *man* in the Divine understanding, such as original sin, could be true of each particular enmattered man (Jones 1969, 188).

The nominalists, by contrast, took the view that universals were mere words, and did not signify any *thing*. Roscelin of Champiègne, often taken to be a leader of this school, argued that universals were mere vocal utterances (Thompson 1995, 409, 419). This presented problems for standard theological commitments of the time. It seemed to imply that God the

Father, the Son, and the Holy Spirit could be three different persons. The term 'God' could be applied to many different particulars not necessarily sharing some single *thing*. Consequently, in 1093 the Council of Soissons ordered Roscelin to repudiate his teachings (Jones 1969, 189). But this doctrine also seemed to imply, counterintuitively and implausibly, that individual humans had nothing more in common than being designated by the universal term 'human.'

Conceptualism is most distinctively identified with Peter Abelard (or Abailard) (1079–1142), who was a student of both the realist William of Champeaux and the nominalist Roscelin, and unsurprisingly argued for a position between these two extremes. On his approach, universals don't signify things in the world. They signify *things* – concepts – in the mind. He began with the idea that universals are words rather than things. They are words that can be "predicated of many," to distinguish them from the particular terms that are predicated only of individual things.

> A *universal* word, however, is one which is apt by its invention to be predicated singly of many, as this noun man which is conjoinable with the particular names of men according to the nature of the subject things on which it is imposed. A *particular* word is one which is predicable of only one, as Socrates when it is taken as the name of only one. (Hyman and Walsh 1983, 177)

The term 'man,' for instance, can be predicated of many individual things, while the particular terms – the names of these individual things – 'Socrates,' 'Plato,' and 'Boethius,' can be predicated only of a single individual thing.

So far this is consistent with the nominalist position, but then Abelard argued for a rational component through the process of abstraction.

> In relation to abstraction it must be known that matter and form always subsist mixed together, but the reason of the mind has this power, that it may now consider matter by itself; it may now turn its attention to form alone; it may now conceive both intermingled. The two first processes, of course, are by abstraction; they abstract something from things conjoined that they may consider its very nature. But the third process is by conjunction. For example, the substance of this man is at once body and animal and man and invested in infinite forms; when I turn my attention to this in the material essence of the substance, after having circumscribed all forms, I have a concept by the process of abstraction. (Hyman and Walsh 1983, 183)

This abstraction consists in the selective attention to some subset of features of an individual thing. Form can be attended to without attention to matter. Each of these features can be attended to separately, or conjoined with other features, either by likeness or by conjunction.

> Nevertheless, perhaps such a conception too could be good which considers things which are conjoined, and conversely. For the conjunction of things as well as the division can be taken two ways. For we say that certain things are conjoined to each other by likeness, as these two men in that they are men or grammarians, and that certain things are conjoined by a kind of apposition and aggregation, as form and matter or wine and water. The conception in question conceives things which are so joined to each other as divided in one manner, in another conjoined. (Hyman and Walsh 1983, 184)

What distinguishes this view from nominalism is that there is a common cause to this conjunction – the nature of the things themselves. Because Socrates and Plato shared certain features in reality, they can be conjoined under a single universal species term, 'man.' And because the forms of Plato and Socrates are also enmattered in reality by the same substance, they are each conjoined in matter and form. The concepts underlying the use of universal terms are grounded on real features of the world, and are not just arbitrary sounds in spoken language or arbitrary marks in written language.

Abelard revisited the universals framework first established by Porphyry – whether universals subsist, whether they are corporeal, and whether they are sensible. He concluded first, that species and genera subsist because "they signify by nomination things truly existent, to wit, the same things as singular nouns." Second, they are both corporeal and incorporeal. They are corporeal "with respect to the nature of things and incorporeal with respect to the manner of signification." Third, they are both sensible and insensible: species and genera "are sensible with respect to the nature of things, and that the same are insensible with respect to the mode of signifying" (Hyman and Walsh 1983, 186–187). The insight here is that the classificatory terms 'species' and 'genus' are associated with mental objects in conception, but they also reflect things of the world in application. This is contrary to the nominalists, who denied the terms were grounded on things in the world and the realists, who seemed to neglect their psychological basis in concepts.

In contrast to the Neo-Platonists, Abelard reconciled Plato and Aristotle by making Plato more Aristotelian. The Neo-Platonists typically reconciled the two by restricting Aristotle's views in *The Categories* and *On Interpretation* to being just about language, preserving the independence of Plato's transcendental forms. But Abelard turned Plato's transcendental forms into concepts of the mind. These concepts exist in some sense independent of things in the world, but they are not existent in some transcendental intelligible realm of forms, independent of all thought. In the century after Abelard, many of Aristotle's works were rediscovered through the translation of Arabic texts. No longer was commentary limited to just *The Categories* and *On Interpretation*, as understood by Porphyry and Boethius. But all this new material upset the traditional interpretation of Aristotle within a Christian, Neo-Platonist framework, and there were attempts to ban Aristotle's work in 1210 and 1215. But by the 1240s, the academic curriculum had absorbed Aristotle's newly discovered works, in spite of the condemnations of them as theologically dangerous in 1277, 1284, and 1286 (Marrone 2003, 35). A new, more comprehensive understanding of Aristotle was now possible, and was undertaken by Thomas Aquinas, who fought the efforts to suppress Aristotle's views by trying to reconcile this new more complex Aristotle with Christian theology. Aquinas was clearly a realist, committed to the idea that universals were ideas in the Divine understanding (Brown 1999).

A seemingly novel understanding of universals was developed by William of Ockham (1280–1349), whose emphasis on language has led many to characterize him as a nominalist. He argued for a hierarchy of language, where written and spoken language are dependent on a mental language. In his *Summa Totius Logicae*, he started with what he took to be the Aristotelian view, as laid out by Boethius:

> According to Boethius in the first book of the *De Interpretatione*, language is threefold: written spoken and conceptual. The last named exists only in the intellect. Correspondingly the term is threefold, viz. the written, the spoken and the conceptual term. A written term is part of a proposition written on some material, and is or can be seen with the bodily eye. A spoken term is part of a proposition uttered with the mouth and able to be heard with the bodily ear. A conceptual term is a mental content or impression which naturally possesses signification or consignification, and which is suited to be part of a mental proposition and to stand for that which it signifies. (Hyman and Walsh 1983, 653)

For Ockham, a universal term, 'man' for instance, may exist as a spoken word or a written word, but it also relies on a mental term. Written and spoken languages are conventional, but mental language is not because mental terms "naturally" signify what they are concepts of - what they are predicated of in nature. By this Ockham meant that a mental term is causally related to particular things in the world, things that are similar in some important way (Normore 1990, 56–57). For Ockham the mental term was not a universal *thing*, whatever that could possibly be, but it was universal in that it naturally signifies many things. This view of Ockham seems to be more like conceptualism than the nominalism of Roscelin that treated universal terms as nothing more than vocalizations. Many historians therefore now see Ockham as a type of conceptualist rather than a nominalist (Hyman and Walsh 1983, 649). Others, however, see him as "nominalist realist," because the natural signification of universal terms is based on real similarities in nature, prior to the activity of a mind. 'Man,' for instance signifies all things that are human, because the mental term grounding this written term signifies naturally all things that are alike in a particular way - the suite of traits shared by all humans, and not just because they are thought alike (Brown 1999, 272).

In the period following Aristotle there seem to have been five different approaches to the understanding of the species and genera of universals. First was the Neo-Platonic realism that identified universals with transcendent forms. Second was the Christian version of Neo-Platonism that identified universals with ideas in the mind of God. Third was the nominalism of Roscelin that treated universals merely as spoken or written words. Fourth was the conceptualism of Abelard that identified universals with concepts of the mind derived by abstraction from the properties of particular things. Finally was the conceptualism of Ockham that identified universals with the naturally signifying terms of a mental language, on which spoken and written language depends.

So what does all this theorizing about the nature of universals have to do with biological classification? These commentators had little interest in the nature and classification of *biological* taxa. They were unaware of, and likely would not have cared about Aristotle's biological uses of the terms 'eidos' and 'genos' as enmattered forms perpetuated in generation, and as developmental articulations. They used the corresponding Latin terms 'species' and 'genus' to refer to linguistic entities - universal terms, rather than to kinds of living things. And the classifications that were generated by the use of these terms were often nonbiological. Boethius, for instance, regarded *justice* as a species of the genus *the good* (Hyman and Walsh 1983, 130).

But no one else of the time and place was really engaged in a specifically *biological* classification either. Perhaps the closest anyone came to doing this before the Renaissance were the authors of the beastiaries and falconry texts. The beastiaries were descriptions of the various kinds of "beasts" – both real and imaginary. They were typically not intended, however, to be accurate or systematic descriptions of animal kinds, but instead to represent and illustrate moral virtues and vices (Ogilvie 2006, 97). The "Cocodryllus" of the Nile, for instance, described as about thirty feet long, with horrible teeth and claws, was also hypocritical and puffed up with pride (Ogilvie 2006, 102). The falconry texts of Frederick II and Albertus Magnus, on the other hand, involved more systematic descriptions of animals. But like the beastiaries, they were largely practical – albeit in support of the practice of falconry rather than moral improvement.

Frederick II (1194–1250), a Holy Roman Emperor who was excommunicated multiple times for challenging the pope over secular power, read and was keenly interested in the recently translated biological works of Aristotle. He wrote a book on falconry, translated as *The Art of Hunting with Birds*, where in his preface he tells us of the errors in Aristotle's biological works, in particular of his description of birds. The problem with Aristotle, according to Frederick II, is that he was ignorant of the practice of falconry, and relied on hearsay (Wilkins 2009, 39–40). But Frederick II didn't quite get Aristotle right. In his discussion of raptorial birds, for instance, he assumed that Aristotle adopted a dichotomous, privative approach to division and classification:

> It was the habit of Aristotle and the philosophers to classify objects into positive and negative groups and to begin their discussions with the positive. Since it is our purpose to give special attention to raptorials, we shall first consider the non-rapacious (or negative) varieties; afterward we shall consider at length raptorial birds. (Wilkins 2009, 40)

But as we have seen, Aristotle rejected this dichotomous classification of living things based on privative terms. Perhaps Frederick II was misled by the early commentaries that were written in ignorance of Aristotle's biological works. Regardless, though, he accepted one of Aristotle's categories, *waterbirds*, but distinguished it from *non-waterbirds*, seemingly for its potential relevance to falconry. Frederick II also promised a treatise on the species and genera into which raptors were to be divided, although this apparently was never produced (Wilkins 2009, 40–41).

Albertus Magnus, or "Albert the Great" (1193–1280), was teacher of Aquinas and one of the first to write commentaries on Aristotle after the rediscovery of his works in the twelfth century. Like Frederick II, he had an interest in falconry, and was likely familiar with Frederick's *The Art of Hunting with Birds*. His major works include one book on animals, *De Animalibus*, and another on plants, *De Vegetabilibus*. According to John Wilkins he "describes, species by species, various animals known" (Wilkins 2009, 43). We should be careful about this claim, however, in that, like the others of his time, he would not have restricted the term 'species' to a fixed level of classification. And it is highly unlikely that he would have, or could have, precisely identified what we would call species today. Like fellow falconer Frederick II, he had a special interest in, and sensitivity to, the various kinds of birds. But he also classified them with bats, presumably based on similar behaviors (Wilkins 2009, 44).

With the beastiaries and the falconry texts we get a somewhat more biological way of thinking about the kinds of living things – in contrast to the linguistic and philosophical approaches of the Aristotle commentators, but the purposes of the authors here were primarily practical – moral edification in the former and falconry in the latter. For the beginnings of a truly *biological* classification we must await the Renaissance.

Conclusion: Two Themes

We began this chapter with the essentialism story: Before Darwin, biological classification was essentialist in that organisms were grouped into timeless and unchanging biological categories based on the possession of a set of timeless and unchanging essential traits or properties. This story located the origins of essentialism in the method of division used first by Plato, and then developed by Aristotle, based on the Greek classificatory terms 'eidos' and 'genos.' This method, according to the essentialism story, was then adopted by all those who followed, up until its demise at the hands of Darwin. One purpose of this chapter is to correct this history, and to better understand the philosophical basis of modern biological classification. What we can say so far is this: First, the essentialism story is misleading – at best – about the views of Aristotle. He used the method of division in his logical works but denied that it could be used to arrive at natural biological kinds. The method of division cross-classifies based on the attributes chosen, it relies on superfluous traits, and it doesn't result in "natural" classifications, which are based on many traits. Second, his nonlogical uses of the terms 'eidos' and 'genos' reveal a complexity in his thinking. In the logical sense these are classificatory terms, where an *eidos* is a specific subdivision of a more general *genos* based on the presence of some differentia. That more general *genos* may in turn be an *eidos* relative to a higher level *genos*. But Aristotle also used these terms historically. An *eidos* is an enmattered form perpetuated in generation. A wading bird, for instance, has a particular appearance or form that is inherited from its parents and passed on to its offspring. *Eide* are not to be thought about just in terms of the features they have, but also in the lineage that produces them. And on a third sense of these terms, an *eidos* is a specific developmental articulation out of a more general foundation – a *genos*. This articulation and differentiation occurs in a functional framework, whereby development properly enhances functioning in an environment relative to a particular lifestyle. Here essences are functional. They are the traits that are necessary for a particular lifestyle in an environment.

But it also seems that the essentialism story got *something* right – at least about the views of the commentators on Aristotle. Their ignorance of Aristotle's rejection of the method of logical division in his biological works, as well as their exclusive focus on language, logic, and metaphysics, perhaps made Aristotle *seem to them* something like the essentialist of the essentialism story. But it isn't clear that even those commentators can really be viewed as essentialists in this narrow way. Their views about language and classification are far too varied, complex, and nuanced to conclude that the species and genera of the Medieval period were taken by *all* to have timeless and unchanging essences.

The realist commentators were the most plausibly essentialist. Whether they thought universal terms – species and genera, signified unchanging Platonic forms in a transcendent realm or timeless ideas in the mind of God, these terms could plausibly be associated with unchanging and timeless essences. On the other hand, the nominalists such as Roscelin thought the relation between words and things to be merely conventional. They could hardly be seen as claiming that universals – *species* and *genera* – signified unchanging natural kinds with essences. Rather, for the nominalists, there were merely definitions of terms as used. Those definitions would surely change as usage changed. Somewhere in the middle was Abelard, with his conceptualism that treated spoken and written language as conventional, while the mental language corresponded to things in the world. According to Abelard, universal terms in spoken and written language – *species* and *genera*, were conventional and as for the nominalists, could not be seen as representing natural kinds with essences. The mental language, concepts in the mind, on the other hand, *might* more plausibly be regarded as essentialist in the manner suggested by the essentialism story. But even that is not so clear.

There is an important theme here that revisits a topic in the first chapter. There we looked at a tension between language, classification, and the world. The general terms that we learn generate classifications, but these classifications do not necessarily "cut nature at its joints." The categories they produce may just reflect practical value or cultural significance. The natural kinds framework has been a standard way to think about this. Natural kinds "cut nature at its joints," while conventional and artificial kinds do not. This captures the idea that some of our classifications seem natural in that they group and divide things to accurately reflect the important groupings and divisions in nature. But as ethnotaxonomy and developmental linguistics reveal, the classifications that spontaneously arise from the use of language within a culture do not necessarily reflect the real groupings and divisions in nature.

This tension is apparent in the history of classification as well, particularly in the thinking of Aristotle. Whether or not *The Categories* is *just* about language and predication, it is at least partly about language and predication. Aristotle was concerned about how we apply general terms, the *eide* and *gene*, to things in the world. And as he noted in his logical works, there is a hierarchy in that *eide* are subdivisions of *gene*. But Aristotle was also aware that there was more to it than just the linguistic and logical elements. How the *eide* and *gene* divided up the world was also important. Returning to Plato's metaphor, did a particular use of a universal term, which generates a category, cut nature at its joints? Aristotle didn't think a simple application of his method of logical division to do so in the biological realm. It did not and could not produce a natural classification.

It may be that Aristotle was conscious of this tension because he was not just a logician and linguistic philosopher, but also an empiricist in the broadest sense. He thought that to understand the world we need to observe that world. But our descriptions of the world in language may or may not accurately represent that world. Individual things – primary substances, can be known in one sense only through sensation. And they are particulars in that each is unique. But to know them in another way, we need to understand them through universals, the language that we use to describe them. This sensitivity to the tension between language and the world is not found in the views of many of those that followed. Augustine, for instance, could be confident of the correspondence of the universals and things in the world, because he was less interested in the observable things around him than in the notion that species and genera were ideas in the mind of God. This is true of the Christian Neo-Platonists in general. But this tension between language and the world could not really be denied, especially as the Renaissance brought renewed attention to observation of the natural world.

3 The Darwinian Pivot

The Return of Empirical Thinking

One premise of this book is that we can better understand modern biological classification through its history. In Chapter 2 we looked at its origins in Aristotle's use of the terms 'eidos' and 'genos,' and saw how these terms had multiple meanings, as logical universals, as enmattered forms perpetuated in generation, and as developmental articulation. We also saw how Aristotle's framework was then transformed in the Latin commentaries about species and genera that followed, with their nearly exclusive focus on language and universals. In this chapter we continue this historical project by looking at biological classification after the transformation of Aristotle, in the thinking of the early medical herbalists and naturalists. We then trace this thinking through to Darwin and his use of an evolutionary framework for classification.

In the Medieval commentaries the understanding of Aristotle was limited by the fact that only his logical works were read and discussed. So his reservations about using the method of division to classify living things were unknown. Moreover, the interests of the commentators were not biological but linguistic in the use of universal terms, metaphysical in the nature of what those terms signify, and theological in the reconciliation of Aristotle with Plato and Christian doctrine. There was at this time little interest in the observation of nature. It would therefore be unsurprising if falconers, such as Frederick II and Albertus Magnus, had a more detailed and careful classification of creatures than those who were interested primarily in language, logic, and morals. The practice of falconry requires a knowledge of nature that depends on careful observation. Similarly, and as we saw in Chapter 1, those who had to rely on their knowledge of nature to live, such as the New Guinea highlanders, typically also had more detailed classifications than those who do not, such as urban college students. Practical needs can surely motivate careful classification. This is what we also see in the first empirically based classifications of the early modern era, by the medical herbalists, who wanted to improve the practice of medicine by improving the preparation of medications. They required a system for reliably identifying the specific plants used in formulating each medication. Given these goals, it is not surprising that the medical herbalists were largely unconcerned with the philosophical debates about universals. And while they typically started with the texts of Aristotle, Theophrastus, Dioscorides, and Galen, they did not just assume that these texts were accurate (Ogilvie 2006, 97–98).

In the earliest group of medical herbalists, the most important figure was Niccolo Leoniceno (1428–1524), who taught moral philosophy and medicine at the University of Ferrara. He had a large library containing many Greek manuscripts, including some of Aristotle's biological works, in both Greek and Latin (Ogilvie 2006, 30–31). He was determined to correct the errors in these texts through the careful observation of nature. A second group of herbalists, many with medical training, included Otto Brunfels (1488–1534), Hieronymous Bock (1498–1554), Leonhart Fuchs (1501–1566), and Peitro Andrea Mattioli (1501–1577) (Ogilvie 2006, 36). Conrad Gessner (1516–1565) was also an important member of this group. Known primarily as a botanist for his *Historiae Plantarum*, which contained descriptions and drawings of more than 1,500 plants, he also produced a massive work on animals, *Historiae Animalium*, written from 1551 to 1558 (Wilkins 2009, 55). This group of herbalists was still practical in orientation, devoted to the improvement of medicine, but they were also more focused on the description of plants and animals in general. This led to an increased emphasis on the gathering of plants for botanical gardens and herbaria (Ogilvie 2006, 39). But as the collections grew larger, the problem of organization became more pressing. *How* should a herbarium or botanical garden be organized? And *why* should it be organized in any particular way?

A third group of medical herbalists started to work out answers to these questions. Descriptions and illustrations were becoming more detailed and precise, but there was still no single naming scheme, so each naturalist could use a different name to refer to a single kind of plant. This problem became more pressing as more and more plants were named. The Lyon herbal of 1587–1588, for instance, included descriptions of more than two thousand plants (Ogilvie 2006, 48). Caspar Bauhin (1560–1624) did his best to correct this disparity in naming schemes. He devoted more than thirty years reading through botanicals, comparing descriptions and names with specimens in his own herbarium, trying to resolve the conflicting names and descriptions. As a result of his work, plant descriptions became more precise and therefore more informative in distinguishing the different kinds of plants (Ogilvie 2006, 48).

At this time many herbaria were organized by alphabetical order. Another approach was on subjective criteria, such as *plants with flowers that please* and *odorous plants*. Sometimes ecological criteria were also used, such as *plants found in shadowy, wet, and damp places* (Ogilvie 2006, 216–217). But some classifications were based on similarity. Hieronymous Bock, Valerius Cordus, and Leonhart Fuchs all tended to group similar plants together (Larson 1971, 9–11). This led to hierarchical and more systematic approaches. Bauhin, for instance, began organizing plants into genera and subdividing them into species. But these species and genera were not necessarily what we would see in modern scientific classifications. Sometimes plants were simply classified as *trees*, *shrubs*, or *herbs* (Ogilvie 2006, 216–219).

Aristotle Returns?

The medical herbalists were more interested in practice than in theory. What was the best way to describe and identify the particular plants required for the various medicines so that the apothecaries would be least prone to error? And what was the best way to organize this information to aid in memorization? This all required a single naming scheme and a user-friendly system of classification for the herbals and herbariums. Andrea Cesalpino (1519–1603), perhaps the most important herbalist of this time, accepted these practical goals, but also adopted a number of what he took to be Aristotelian commitments. First, he agreed with Aristotle's epistemology, where the highest form of knowledge consists in the predication of universals (Larson 1971, 24, 38). We must know what universals apply to a thing to truly understand it. In other words, we need to know which species and genus terms can be predicated of a thing. Second, he focused on functionally essential traits, the nutritive and reproductive traits required for life. Third, in agreement with the Medieval misunderstanding of

Aristotle, he used the method of division, based on functionally essential traits associated with nutrition and reproduction, to classify living things (Larson 1971, 27).

As his use of the method of division implies, Cesalpino's approach was hierarchical, but it was a hierarchy only of species and genera, not of the kingdoms, classes, and orders of the Linnaean system (Ogilvie 2006, 224). A genus at one level of analysis could be a species at another. The first development of a fixed hierarchy is seen in the work of the systematists who followed and built upon Cesalpino's framework: Robert Morison, Augustus Rivinus, Josef Pitton de Tournefort, and especially John Ray (Larson 1971). This group continued to use 'species' and 'genus' as the primary classificatory terms, but there was also an increasing tendency to identify fixed taxonomic levels. John Ray (1627–1705), for instance, began to use the term 'genus' (sometimes 'order') only for the most general taxa. The term 'species subalternae' (sometimes 'tribe') designated an intermediate level, and the term 'species infimae' the lowest, least general level (Ray 1735, 21). Like his predecessors, Ray also used 'species' and 'genus' to refer to nonbiological categories. He speculated, for instance, about whether all metals were of the same species or not, and the number of species of indivisible particles or atoms (Ray 1735, 60–61).

Cesalpino, and many of those who followed, relied on those traits identified by Aristotle as *functionally* essential traits to classify – reproductive traits in particular. But because they used the method of division based on just a few traits, they also used these traits as *taxonomically* essential to determine inclusion in a category. As we have seen, this ran counter to Aristotle's views, when he argued that classification should be based on many traits. But the traits used were also typically practical in the sense that they were highly variable and easy to observe and identify. This use of different criteria often did not produce any conflict, as many reproductive traits are also highly variable and easy to identify and observe. But there was nonetheless a tension here. Sometimes practical demands prevailed and species and genus groupings were based just on the highly variable and easily identifiable "accidental" traits, rather than the functionally essential traits. For this reason, these naturalists typically admitted that their classifications were artificial rather than natural. For them, an artificial classification was based on accidental traits while a natural classification was based on functionally essential traits.

That was one sense of the terms 'artificial' and 'natural' as they applied to classification. But there was another sense as well, in which a natural classification would be based on *all* similarities and differences, and would reflect the continuities in nature, as James Larson explained:

> In relation to plant form, the natural method, when perfected, would admit no exceptions, and would be independent of the practical interest or value imputed by human will. Comprehending all plant parts, properties, faculties and qualities, such a method would consider roots, stems, and leaves, as well as flowers and fruits, and draw from the comparison of their resemblances and differences the affinities which resolved them in groups. (Larson 1971, 42)

This view of what makes a classification natural echoes Aristotle's rejection of the method of division to classify living things. As we saw in the last chapter, he had argued that classification should be based on many traits and reflect the full complexity and continuity in nature.

There is an important point here. Given the context, it is not clear why a classification should be based on all traits rather than on just a few functionally essential ones. In fact, except for the purely practical criteria such as easy identification and high variability, it is unclear why *any* particular trait would be better than any other for classification. We can understand this problem relative to modern evolutionary thinking about classification. On the now standard approach, biological classification represents phylogeny and the evolutionary tree. This theoretical basis gives criteria for determining which traits should be used in classification. As we shall see, *homologies* – shared traits based on common ancestry – are better traits for generating a classification on this theoretical basis. There is a theory, evolutionary theory, that gives us operational guidance here about which traits to use. But it isn't clear what *theory* these pre-Darwinian naturalists had that could guide them here. The assumption that God created all living things according to some plan cannot by itself give guidance on this question, unless God's plan could be known independently. We might think perhaps that God intended living things to flourish, and therefore we should use traits functionally essential for life. But alternatively we might think that a classification based on all traits would better express the nuances in God's intentions. It isn't obvious how these early naturalists could reconcile these conflicting tendencies.

Language and Error

During this period when a practical system of biological classification was being developed by the medical herbalists and the early naturalists such as Cesalpino and Ray, there was also an ongoing philosophical debate about language and classification that continued along lines similar to those of the Medieval discussions of universals. But at this time, the focus turned as well to epistemology. How can we have knowledge and avoid error? Most notably, this was Rene Descartes' (1596–1650) starting point in his method of doubt. But we also see this turn to epistemology in the work of Francis Bacon (1561–1626) and John Locke (1632–1704). Neither of these thinkers is known for his views on biological classification, but they are nonetheless philosophically relevant to the history of biological classification.

According to Bacon the mind is like a mirror that reflects things in the world, but one that has distortions that need to be eliminated or corrected. He called these distorted images the "idols of the mind," because although they may be venerated they also misrepresent the Divine Ideas that guided creation (Bacon Bk. I, XLII). According to Bacon, there are four idols of the mind. The *idols of the tribe* are due to the inherent tendencies for error in common human nature – through rush to judgment, imagination, and exaggeration. The *idols of the cave* are due to individual tendencies for error – through temperament, education, habit, and environment. The *idols of the theatre* are due to sophistry and false learning – through the false theories of philosophers and theologians. The *idols of the marketplace* are due to the false significance bestowed upon words – through the demands of communication and the careless use of language. According to Bacon, these *idols of the marketplace* are the most troublesome of all:

> For men believe that their reason governs words; but it is also true that words react on the understanding; and this it is that has rendered philosophy and the sciences sophistical and inactive. Now words, being commonly framed and applied according to the capacity of the vulgar, follow those lines of division which are most obvious to the vulgar understanding. And whenever an understanding of greater acuteness or a more diligent observation would alter those lines to suit the true divisions of nature, words stand in the way and resist change. (Bacon 1960, Bk. I, LIX)

The errors here are of two kinds. The first error is using words as names for things that do not exist. Examples of this kind of error, according to Bacon, include the words 'fortune,' 'prime mover,' and 'planetary orbits.' Common use of these terms makes us think incorrectly that these things exist. The second kind of error is using words as the names of things that exist, but the ideas associated with the names are "confused and ill-defined, and hastily and irregularly derived from realities." According to Bacon, this second kind of error is exemplified by terms that refer to qualities such as 'humid,' 'heavy,' 'light,' 'rare,' and 'dense.' Common use here produces confused and ill-defined ideas (Bacon 1960, Bk. I, LX). We can see this in modern usage. The terms 'heavy' and 'light,' for instance, are used in a variety of ways, and with little precision. We might have heavy thoughts, or be carrying a heavy load, and what is a heavy load for one person may be light for another.

Although Bacon didn't think the terms that refer to biological categories are normally misleading in these ways, he thought there was still potential for confusion:

> Our notions of less general species, as Man, Dog, Dove, and of the immediate perceptions of the sense, as Hot, Cold, Black, White, do not materially mislead us; yet even these are sometimes confused by the flux and alteration of matter and the mixing of one thing with another. (Bacon 1960, Bk. I, XVI)

Bacon does not explain precisely what this "flux and alteration of matter and the mixing of one thing with another" might involve. But the solution to this idol, and the others, is careful, unbiased observation, followed by generalization and classification. This required, according to Bacon, that we shun mere hypotheses or "anticipations of nature." Whether or not this is plausible (and we will return to it in Chapter 8), Bacon was surely right in his concern that the words we use do not always divide nature at its joints, and that error can therefore arise from the everyday, vulgar use of these words. The point is this: In learning words in a language, the categories we acquire may or may not divide the world naturally. We saw this in Chapter 1, in the distinction between natural and mere conventional or artificial kinds. We can learn and apply the word 'weed,' for instance, but that fact does not make *weed* a real, natural category.

We see a similar concern in the writings of John Locke a century later. Like Bacon, he saw the use of language as a source of error, as words are "doubtful and uncertain in their significations." Locke began by noting that to communicate, we need to use general terms that signify multiple things and sort things into groups. In part this is because we simply could not form distinct ideas of each individual thing – every bird and beast, tree and plant (Locke 1975, Bk. III, Ch. III, §1–3). So we form general abstract ideas, signified by universal terms that can be used to refer to these individual distinct things, but not *as* individual things. We get the general idea of *horse*, for example, by observing all the particular instances of horses, eliminating what differs among these instances and keeping what is common. Similarly, we get the general idea of *animal* by looking at all the instances of animals – individual horses, humans, cats and dogs, and so forth, and retaining only what is common. Through further abstraction we also get the ideas of *body* and *life*. According to Locke, this process of abstraction is how we get our species and genera:

> To conclude, this whole mystery of genera and species, which make such a noise in the schools, and are with justice so little regarded out of them, is nothing else but abstract ideas, more or less comprehensive, with names annexed to them. (Locke 1975, Book III, Ch. III, §9)

Abstract ideas are defined, according to Locke, by giving the genus and the differentia that distinguishes that species within the genus. These abstract ideas then form the basis for our categories. The abstract idea of *animal*, for instance, gives meaning to the term 'animal' and serves as the foundation for the category of *animal* (Locke 1975, Bk. III, Ch. III, §10).

Locke thought this process was also the basis for knowledge. Echoing Aristotle, Locke claimed that higher knowledge requires more than just acquaintance with particular things through sensation, but also the generalizations and categories implicit in universal terms (Locke 1975, Bk. III Ch. III, §4). The reason we classify into categories – species and genera, according to Locke, is because we naturally desire knowledge:

> For the natural tendency of mind being towards knowledge; and finding that, if it should proceed by and dwell upon only particular things, its progress would be very slow and its work endless; therefore, to shorten its way to knowledge, and make perception more comprehensive, the first thing it does, as the foundation of the easier enlarging its knowledge, either by contemplation of the things themselves that it would know, or conference with others about them, is to bind them into bundles, and rank

> them so into sorts, that what knowledge it gets of any of them may thereby with assurance extend to all of that sort; and so advance by large steps in that which is its great business, knowledge. This, as I have elsewhere shown, is the reason why we collect things under comprehensive ideas, with names annexed to them, into genera and species; i.e. into kinds and sorts. (Locke 1975, Bk. II, Ch. XXXII, §6)

But this process can also go wrong. Words *seem* to "carry in them knowledge of their essences." For Locke, the essence associated with the word may not correspond to the true essence of the thing though, because the vulgar use of terms is not based on the true nature of things. In other words, the meaning of terms may not be a good guide to the true nature of the things they signify.

This worry is behind Locke's distinction between *nominal* and *real* essences. The *nominal essence* is the abstract idea to which a name is attached (Locke 1975, Bk. III, Ch. VI, §2). The *nominal essence* of gold, for instance, is the complex idea that the word 'gold' stand for "a body yellow, of certain weight, malleable, fusible and fixed." In contrast, the *real essence* is the underlying constitution of gold, on which the other properties depend. In modern chemical theory, we would say that the real essence of gold is its atomic number (79), which is responsible for many of its other properties. The nominal essence of a term, then, is its meaning as established by abstraction based on observation, experience, and usage, and the real essence is the underlying nature of the thing designated by the term.

The problem with language, according to Locke, is that the meanings of our general terms get determined by vulgar use, based on everyday experience. And when we think of something based on this *nominal essence*, we are not thinking of it in terms of its true nature – its *real essence*. Locke was skeptical that we could ever get at the true natures of many things, such as the substance gold, because of our perceptual limits. To understand the real essence of gold, for instance, we need to do more than just look at rings and coins. Modern chemical theory tells us that we need to observe it at the atomic level. Locke was understandably skeptical that we could ever know real essences.

Gottfried Leibniz (1646–1716) was more optimistic than Locke. He recognized that although the words we use predate scientific investigation, and therefore reflect earlier states of ignorance, language also gets corrected in the process of science (Leibniz 1996, 319; see also Wilkins 2009, 64–65).

This is an important point for understanding the relation between vulgar language and its influence on classification. We may start with the general terms used to communicate, terms that may falsely imply the reality of some things or that may give us confused notions about things that exist (as Bacon claimed), but through careful investigation we can correct these ways of thinking about the world and thereby correct the ways we categorize and classify things in the world. Perhaps this is what happened in the development of an evolutionary framework for classification in the period from Linnaeus to Darwin.

The Linnaean System

The Swedish naturalist Carolus Linnaeus (1709–1778), born just after the death of John Ray in 1705, is a central figure in modern biological classification. He adopted and developed the basic approach proposed by Cesalpino and Ray. From Ray, he took the fixed hierarchy of three levels and expanded it to five levels, by adding the ranks *order*, *class*, and *kingdom*. In doing so, he restricted the use of the term 'species' to the lowest level. No longer was this term, along with 'genus,' applied at multiple levels, depending on the inquiry. And following Cesalpino's example, Linnaeus used the functionally essential sexual or fructification traits to group organisms. But like all of his predecessors, he also used practical traits that were easily identifiable and highly variable. And as is well known, he also proposed and used the familiar binomial nomenclature that gives species taxa a genus–species name, as in *Homo sapiens*. But more than any of his predecessors, he also took a theoretical interest in classification. What is the nature of biological taxa, what should a classification represent, and which characters or traits should we use to classify?

According to the essentialism story sketched out at the beginning of Chapter 2, Linnaeus was an arch essentialist, using Aristotle's method of logical division to classify organisms on the basis of taxonomically essential – necessary and sufficient – traits. This is usually assumed to imply that Linnaeus believed species and higher-level taxa to be unchanging, eternal, and discrete. This interpretation of Linnaeus is certainly understandable. In the first edition of his main classificatory work, the *Systema Naturae*, it certainly looks like he is doing something like this. But a careful look at his career paints a very different picture.

In 1735, Linnaeus left Sweden for the Netherlands to earn a degree in medicine. That year he published his first major work, the *Systema Naturae Sive Regna Tria Naturae Systematice Proposita per Classes, Ordines, Genera & Species.* In this fourteen-page first edition, he laid out the framework for his system. There were three kingdoms of nature - plant, animal, and mineral, each subdivided into nesting classes, orders, genera, and species. Here he also began his theoretical speculations on the nature of biological taxa. In the first of his "Observations on the Three Kingdoms of Nature," he claimed that no new species are ever produced: "Hinc nullae species novae hodienum producuntur" (Linnaeus 1964). In the fourth observation, he argued for the fixity of species on that ground that since offspring closely resemble parents: "like always gives birth to like."

His views changed seven years later, when he observed what he thought was a specimen of the flower *Linaria*, which had some of the distinct attributes of another species, *Peloria*. As the offspring of this specimen were themselves fertile, he was convinced that a new species and new genus had been generated through hybridization. Linnaeus took this to be evidence of change, not just of degree but also of kind (Larson 1968, 293; Eriksson 1983, 94–95). He defended this claim in his *Dissertatio botanic de Peloria*, and later described his conclusions in a letter to Albrecht von Haller:

> I beg of you not to suppose it [the *Peloria*] anything else than the offspring of *(Antirrhinum) Linaria*, which plant I know well. This new plant propagates itself by its own seed, and is therefore a new species, not existing from the beginning of the world; it is a new genus, never in being until now. (Larson 1968, 294)

Linnaeus concluded that new species were formed through hybridization, contradicting his claim in the first edition of the *Systema Naturae* that no new species are ever produced. He came to think this to be relatively common. This conclusion was incorporated into later editions of his *Systema Naturae*, along with a theological explanation. In the tenth edition of 1758, he wrote that God created an original individual or mating pair for each genus, and new species were produced by intergeneric crosses. And in the thirteenth edition, published in 1770, he speculated that the original individual or breeding pair might instead represent *orders* rather than genera, and that new genera, as well as new species, could be formed through hybridization.

> We may suppose God at the beginning to have proceeded from simple to compound, from few to many! and therefore at the beginning of vegetation to have created just so many different plants, as there are natural *orders*. That He then so intermixed the plants of these orders by their marriages with each other; that as many plants were produced as there are now distinct *genera*. That Nature then intermixed these generic plants by reciprocal marriages ... and multiplied them into all possible existing species. (Larson 1968, 297)

Significantly, he did not say that there were limits to this creative process (Larson 1971, 107).

Linnaeus did not just speculate that hybridization might produce new species; he also proposed a possible mechanism in his *Sponsalia plantarum* of 1746. By 1760, in his *De Sexu plantarum*, he had worked out the details of his theory. Here he speculated that there were two main elements in reproduction: the female "medulla" that was the "bearer of life," and the male "cortex" that was the "conveyer of nourishment" (Larson 1971, Lindroth 1983, 44). The medulla was more "essential" and less variable than the cortex. While this implies that the offspring will resemble the parents in some vague manner or other, it does not imply that they will be identical to the parents. The traits of the offspring will be some mix between the more essential medulla and the more variable cortex. And if the medulla and cortex come from different species or genera, then we should expect the range of variation to be even greater. Significantly, Linnaeus saw this as a natural process governed by law, and with no known limits of change (Larson 1968, 297–298).

What is striking here is that, to begin with, this is not the Linnaeus of the essentialism story, who assumed biological taxa have the unchanging, timeless essences established by Aristotle's method of division. Instead, after 1742, Linnaeus believed species, genera, and orders to be changing and dynamic. There were, however, two ways he *might* be regarded as an essentialist. First, he might be regarded as a *genealogical* essentialist, in that he took membership in a biological taxon to be determined by placement in a genealogy or lineage that ultimately stretched back to the creation. Linnaeus thought about biological classification historically. Second, he *might* be regarded as an essentialist in the manner laid out by Locke, in his distinction between nominal essences and real essences. For Locke, real essences were the underlying structures that made something what it was.

Similarly, for Linnaeus the inherited medulla made a plant what it was, albeit with influence from the cortex. This is not a definitional or taxonomic essence based on the method of division, but a *physical* essence based on the natures of the medulla and cortex.

But if Linnaeus was not a taxonomic essentialist who defined biological taxa with a set of essential traits, how should we understand his *Systema Naturae*, where he seemingly classifies organisms on just one or a few traits - as we might expect if he were using Aristotle's method of logical division? The answer to this question is found in Linnaeus's conflicted understanding of his project. He believed that in studying and classifying life, he was fulfilling a duty to understand the Divine understanding in the creation. For Linnaeus, a truly natural classification was one that represented all living things, in all their complexity, as conceived by God at creation. He believed that he was far from achieving a natural classification, in part because many plants and animals were simply unknown, and therefore could not be placed properly in the classification (Larson 1971, 70). But also a natural classification would be based on all traits, all similarities and differences. Only then would God's plan be fully revealed. The functionally essential fructification traits he used to classify were only provisional - a substitute for the natural method that might ultimately lead to a truly natural classification (Lindroth 1983, 21).

There were several tensions in Linnaeus's views. First was the longstanding tension between practice and theory. From the medical herbalists on, it was important to have a method of classification that was easy to apply in the identification of plants and animals, yet the taxa must also be easy to remember. This meant that the most variable and easily observable traits, such as the number of stamens, had the most practical value in generating classification, precisely because they varied significantly from one type of organism to another and were obvious. But these were not necessarily the most theoretically significant traits. From Cesalpino on, functionally essential traits associated with reproduction and nutrition somehow seemed more important because they were necessary for life. Surely a classification based on these important traits would be more "natural" than one based on trivial traits unimportant for life.

The second tension was between the logical - the simplifying use of just a few traits to generate a fixed, unambiguous hierarchy, and the observational - the multitude of observed similarities, differences, and gradations among living things that seemed to divide and group in many conflicting ways, and without clear lines of demarcation. The goal, according to Linnaeus, and echoing Aristotle, was a natural classification based on all features, not just a functionally essential few. Only then could a classification fully represent the ideas in the mind of God that guided creation. Linnaeus believed that his system was artificial in this way, misrepresenting nature and the mind that created it, but it was nonetheless useful as a start toward a natural classification.

The third tension was between the use of similarities to classify and the historical conception of biological taxa. Linnaeus classified organisms into species, and species into higher-level taxa based on similarities, largely in sexual traits. But he also thought of biological taxa as historical entities – reproductive lineages extending back to creation. But even though "like produces like," it wasn't clear that all similarities were good guides to genealogy. For Linnaeus there was a gap between practice and theory.

Buffon versus Linnaeus

This tension between the oversimplifying, logical system of classification and the observed complexity of nature was also found in the thinking of Linnaeus's contemporary and rival, George-Louis Leclerc, the Comte de Buffon. (1707–1788) But where Linnaeus seemed inclined more toward the side of simplification – even if only provisionally, Buffon seemed to be inclined to the other side, favoring the observed complexity of nature. He laid out his views in the thirty-six volumes of his *Histoire Naturelle*, written over a forty-year period from 1749 to 1788. In the first volume he criticized the arbitrariness of Linnaean classification on the grounds that there were no clear gaps or joints in nature, but the Linnaean system made these cuts anyway, and on the basis of some single arbitrarily chosen trait. If we instead used the many traits we observe, there would be many plants or animals classified in multiple different categories. No single, universal classification, such as the one proposed by Linnaeus, would be possible. According to Buffon, ultimately only individual organisms exist; the higher-level taxa are figments of imagination:

> In general, the more one increases the number of one's divisions, in the case of natural products, the nearer one comes to the truth; since in reality

> individuals alone exist in nature, while genera, orders, classes, exist only in our imagination. (Lovejoy 1968, 90)

But species seemed to be different for Buffon. In the second volume of his *Histoire Naturelle*, published in the same year as the first, he seemed to allow for some reality to species based on a reproductive criterion:

> We should regard two animals as belonging to the same species, if, by means of copulation, they can perpetuate themselves and preserve the likeness of species; and we should regard them as belonging to different species if they are incapable of producing progeny by the same means. Thus the fox will be known to be a different species from the dog, if it proves to be a fact that from the mating of a male and female of these two kinds of animals no offspring is born; and even if there should result a hybrid offspring, a sort of mule, this would suffice to prove that fox and dog are not of the same species – inasmuch as this mule would be sterile. (Lovejoy 1968, 93)

Here, Buffon seems to have claimed that species are groups of organisms that through sexual reproduction can produce fertile offspring. But if there was some process, such as the reproduction of fertile offspring, binding organisms into a species, then that nature can't consist *just* in individual organisms. Species groupings would not be mere figments of the imagination.

Setting aside this apparent contradiction, there is something important here. Buffon wasn't arguing for the classification of organisms into species on the basis of a few traits or attributes, or even many. He was proposing that the ability to interbreed and produce fertile offspring was the relevant criterion. This reproductive species criterion allowed for the possibility of change. In his fifth edition of the *Histoire Naturelle*, Buffon argued that significant change had in fact happened:

> Though Nature appears always the same, she passes nevertheless through a constant movement of successive variations, of sensible alterations; she lends herself to new combinations, to mutations of matter and form, so that today she is quite different from what she was at the beginning or even at later periods. (Lovejoy 1968, 104)

This change was due to the environment, and it was gradual, according to Buffon in his sixth volume of 1756:

> If we consider each species in the different climates which it inhabits, we shall find perceptible varieties as regards size and form; they all derive an impress to a greater or less extent from the climate in which they live. These changes are made slowly and imperceptibly. Nature's great workman is Time. He marches ever with an even pace, and does nothing by leaps and bounds, but by degrees, gradations and succession he does all these things; and the changes which he works - at first imperceptible - become little by little perceptible, and show themselves eventually in results about which there can be no mistake. (Lovejoy 1968, 104)

There may be good reasons to doubt that Buffon thought that this change could be unlimited (Eddy 1994), but whether or not there were limits, this process produced a complex mix of gradations in multiple traits that could not be captured by any abstract, simple system.

Buffon's views are notable also for their historical emphasis. He thought of species not just in reproductive but also historical terms - as *lineages*, as he proclaimed in 1749: "This power of producing its like, the chain of successive existences of individuals, constitutes the real existence of the species [*existence réele*]." A few years later, in 1753, he made a similar claim: "It is neither the number, nor the collection of similar individuals which makes the species. It is the constant succession and uninterrupted renewal of these individuals which constitute it." But notably, he also extended this historical way of thinking to higher-level taxa - to genera and families, as groups of interconnected lineages, and to lower levels - subspecies, as subordinated to species lineages (Sloan 1979, 117–118). This is a natural extension of the view that species are historical entities, because after all, to the degree that there are things like genera, orders, classes, and kingdoms, they are ultimately constituted by species lineages.

The Linnaean system, according to Buffon, was not simply a provisional attempt to uncover the secrets of nature, as Linnaeus saw it; it was a misleading and counterproductive abstraction imposed on nature by the mind (Sloan 1976, 359). True knowledge, according to Buffon, comes not from abstract and logical systems, but is found in the observation of real order in nature. There is an important issue lurking behind the Buffon–Linnaeus dispute. How can an abstract system - a model - represent the world? Models in science generally *mis*represent the world through simplification. We can, for instance, model the gravitational attraction of the Earth and moon in a simple abstract system containing two point masses. But this obviously misrepresents reality. There are many other masses that would have some effect – most obviously the Sun and other nearby planets, but also to some degree, all the masses in the universe. Moreover, none of these bodies are point masses, but each has mass distributed unevenly throughout a body occupying space.

Similarly, we can model biodiversity as a Linnaean hierarchy, with clear and distinct divisions and groupings, logically ordered, and based on a few abstract ideas of traits to make the groupings. In reality though, each organism is unique, and the groupings and divisions of the organisms are not clear and distinct. There are many traits that vary and grade into each other. An abstract model might be good enough for specific purposes, insofar as it represents the physical system to some degree of accuracy. For many purposes, for instance, the Earth–moon, point– mass model may be close enough to the real system – even though it neglects the attractive forces of all other masses in the universe. But the problem with biological classification in the eighteenth century is that it wasn't clear that it was representing anything physical to which it could be compared. If, as Linnaeus thought, his system of classification was to represent ideas in the mind of God at creation, how could we possibly make a comparison at all, unless we have independent access to those Divine Ideas? This is in contrast to the Newtonian Earth–moon model, where we know more or less what is missing, and the degree to which it therefore falsely represents the world.

Buffon's philosophical commitment to the use of all characters in classification was echoed in some systematists and naturalists who followed. For instance, Michel Adanson (1727–1806), a botanist who worked for a time at the Jardin des Plantes in Paris, constructed a classification of plants consisting of fifty-eight "natural" families, published in the second volume of his *Familles des Plantes* (1763). He explicitly considered as many character traits as possible in generating this classification. For Adanson, a natural classification would be based on all character traits, rather than a select few, which would produce an artificial system (Winsor 2001, 2–4).

Two other prominent naturalists in the French-speaking world, Antoine Laurent de Jussieu (1748–1836), a professor of botany at the Jardin des Plantes, and the Swiss botanist Augustin Pyramus de Candolle (1778–1841), also developed systems based on the use of as many characters as possible. These two disagreed, however, in that de Jussieu believed in continuity among living forms, while de Candolle thought that living forms were separated by gaps (Stevens 1994). The belief in continuity, which we saw in Aristotle and Buffon, leads to the worry that higher-level groupings may be arbitrary. If there are no real gaps in nature, then any cuts must be made on the basis of subjective considerations, human propensities, or practical factors. But for both of these naturalists, a natural classification was nonetheless one that reflected all similarities and differences. An artificial classification was based on just one or a few traits, perhaps for subjective, practical reasons such as ease of observation.

Implicit in the views of both Linnaeus and Buffon was the idea that biological taxa are historical entities – reproductive lineages. This idea was also adopted to some degree by Denis Diderot (1713–1784) and Georges Cuvier (1769–1832) (Sloan 2003, 235). Cuvier went so far as to deny that similarity was a criterion for species at all (emphasis added):

> We imagine that a species is the total descendence of the first couple created by God, almost as all men are represented as the children of Adam and Eve. What means have we, at this time, to rediscover the path of this genealogy? *It is assuredly not in structural resemblance.* There remains in reality only reproduction and I maintain that this is the sole certain and ineffable character for the recognition of species. (Mayr 1982, 257)

The philosopher Immanuel Kant (1724–1804) similarly adopted a reproductive and historical way of thinking about biological taxa, citing the views of Buffon:

> In the animal kingdom the division of nature into genera [*Gattungen*] and species [*Arten*] is grounded on the general law of reproduction, and the unity of genus is nothing else than the unity of the generative force, which is considered as generally active for a determined manifold of animals. Thus, the Buffonian rule – that animals which can generate fertile young and which might show differences in form [*Gestalt*], belong to one and the same physical genus [*Physichen Gattung*] – can properly be applied only as a definition of a natural genus [*Naturgattung*] of animals generally, to differentiate it from all logical genera [*Schulgattungen*]. (Sloan 1979, 127)

As the last sentence here indicates, Kant distinguished the logical system based on similarities from the natural system based on genealogy and reproduction.

> The logical division [of taxonomists] proceeds by classes according to similarities; the natural division considers them according to the stem [*Stämme*], and divides animals according to genealogy, and with reference to reproduction. One produces an arbitrary system for memory, the other a natural system for the understanding [*Verstand*]. The first has only the intention of bringing creation under titles; the second intends to bring it under laws. (Sloan 1979, 128)

For Kant, the historical approach was clearly the better of the two. Shortly thereafter, two other figures followed Kant's example. Christoph Girtanner (1760–1800) and Johann Karl Illiger each endorsed a historical, genealogical conception of species and argued for a reproductive criterion (Sloan 1979, 140–143).

The Nineteenth Century

The next century brought yet another philosophical approach to classification. In 1843, John Stuart Mill published a book on logic with the unwieldy title *A System of Logic: Ratiocinative and Inductive Being a Connected View of the Principles of Evidence and the Methods of Scientific Investigation*. This influential book discussed classification in general, and biological classification in particular, from an empirical, nominalist, and pragmatic stance. Mill began with the idea that classification is a product of our use of language. According to Mill, there are no essences to bind organisms into groups. Rather, organisms are simply bound by a name, based on the presence of certain attributes (Mill 1843, 160). Mill claimed that *which* attributes bind together things into a classificatory category depends on the purpose of the classification.

> And here, to prevent the notion of differentia from being restricted within too narrow limits, it is necessary to remark, that a species, even as referred to the same genus, will not always have the same differentia, but a different one, according to the principle and purpose which preside over the particular classification. For example, a naturalist surveys the various kinds of animals, and looks out for the classification of them

> most in accordance with the order in which, for zoological purposes, it is desirable that his ideas should arrange themselves. With this view he finds it advisable that one of his fundamental divisions should be into warm-blooded and cold-blooded animals; or into animals which breathe with lungs and those which breathe with gills; or into carnivorous, and frugivorous or graminivorous; or into those which walk on the flat part and those which walk on the extremity of the foot, a distinction into which two of Cuvier's families are founded. In doing this, the naturalist creates as many new classes, which are by no means those to which the individual animal is familiarly and spontaneously referred; nor should we ever think of assigning to them so prominent a position in our arrangement of the animal kingdom, unless for a preconcerted purpose of scientific convenience. And to the liberty of doing this there is no limit. (Mill 1843, 174–175)

According to Mill, because these classifications are always devised to suit some purpose or other, they are always artificial in some sense. They are constructed by humans for human purposes, even if based on real clusters of character traits in nature.

But Mill also thought that not all classifications were equal. Some were based on causal traits or properties that could support inductive generalizations. In the 1858 edition of his *System of Logic*, he made this clear:

> The ends of scientific classification are best answered, when the objects are formed into groups respecting which a greater number of general propositions can be made, and those propositions more important, than could be made respecting any other groups into which the same things could be distributed. The properties, therefore, according to which objects are classified, should, if possible, be those which are the causes of many other properties; or, at any rate, which are sure marks of them. (Mill 1858, 434)

A good grouping, in this sense, allows the inference of many other similarities. The grouping of humans into *Homo sapiens*, for instance, allows the inference of many other traits. If one human has some particular trait, the use of language for instance, we can infer that other humans have this trait. However, the important causal properties underlying these regularities are often not "fitted to serve also as diagnostic of the class," so Mill conceded the use of "some of the more prominent effects, which may serve as marks of the other effects and of the cause" (Mill 1858, 434). In other words, and if we were to use Locke's term, the *real essences* are not always appropriate for classification, and we must rely on the effects of these essences. But whatever the case, classifications are "natural" insofar as they support inductive generalizations. Mill criticized the Linnaean system because it was not natural in this sense. The use of fructification or sexual traits did not support generalizations:

> The Linnaean arrangement answers the purpose of making us think together of all those kinds of plants which possess the same number of stamens and pistils; but to think of them in that manner is of little use, since we seldom have any thing to affirm in common of the plants which have a given number of stamens and pistils. (Mill 1858, 433–434)

He concluded that the only value of the Linnaean system is in helping us remember how many stamens and pistils a particular kind of plant may have.

Mill did not deny however, some value to artificial classifications, such as how a farmer might divide plants into *useful plants* and *weeds*. Similarly, whales may or may not be fish, depending on the purposes of the classification.

> If we are speaking of the internal structure and physiology of the animal, we must not call them fish; for in these respects they deviate widely from fishes; they have warm blood, and produce and suckle their young as land quadrupeds do. Though this would not prevent our speaking of the *whale fishery*, and calling such animals fish on all occasions connected with this employment; for the relations thus arising depend upon the animal's living in the water, and being caught in a manner similar to other fishes. (Mill 1858, 436)

So in a natural classification, which supports inductive generalization, whales are clearly not fish, but in an artificial classification, based on the practice of fishing, they might be.

So far, we have looked at four general approaches to classification, based on different views about the purpose of classification, its theoretical basis, and operational method. What makes a classification *natural* depends on purpose and theoretical basis. On the Linnaean approach, for instance, the purpose of classification was to represent the ideas in the mind of God that governed creation. A natural classification would do this by reflecting and representing the ideas of God employed in the creation. The artificial classification that Linnaeus constructed on the bases of one or a few sexual and fructification traits, was merely provisional and partly practical, based on the high variability and observability of these traits. Buffon, Adanson, Jussieu, and de Candolle, by contrast, thought that classification should reflect all the observable similarities and differences across living nature. A natural classification here was simply the one that best reflected *all* similarities and differences. Consequently all traits, not just the sexual, should be used for classification. Mill advocated a pragmatic and epistemic approach. Classifications were intended to represent nature indirectly, through the generation of categories that allowed generalizations. Classifications were epistemically useful things that could serve as a foundation for our inferences about the world. For Mill, natural classifications did just that. And for those who were thinking historically, such as Kant and Girtanner, a natural classification instead correctly represented genealogy and reproduction - grouping organisms together that shared genealogy. It was artificial if it were instead based on mere similarity.

Two more distinctly biological systems were also developed in the early decades of the nineteenth century. The first, by Georges Cuvier (1769–1832) in his *Regne Animal* (1817), was based on a system of four "embranchments" - Vertebrata, Mollusca, Articulata, and Radiata. Each branch was subdivided into four more branches. Cuvier's system was premised on the idea that there were four basic animal body plans, each associated with a particular kind of nervous system. Cuvier also assumed first, that traits should generally be understood functionally, and each individual organism was a functional whole; second, that these traits were responses to the necessities of life within each environment. So to some degree, the patterns of similarity in nature are then due to similarities in conditions of existence (Ospovat 1981, 115; Stevens 1994, 65–72). Biological classification could then be understood to represent solutions to the functional demands of environments.

Four years later, William Sharp MacLeay (1792–1865) published the second part of his *Horae Entomologicae* (1821), in which he sketched out his "Quinarian" system of classification. In this system, animals, insects, and plants were to be arranged into a grand circle of five principal forms. In the animal circle there were five classes - Acrita, Radiata, Annulosa, Vertebrata, and Mollusca. Each of these classes contained five orders, which in turn each contained five tribes. Vertebrata, for instance, contained Reptilia, Aves, Amphibia, Mammalia, and Pisces, each divided into five smaller groups (Hull 1990, 93).

MacLeay also proposed an operational method for generating a classification based on his distinction between *affinities* and *analogies*. Affinities were patterns of many similarities – "agreement among many characters" – and were to be the basis of classification. Organisms in Vertebrata, for instance, shared many similarities based on, among other things, skeletal structures and elements. These affinities were the basis for the circles. Analogies, by contrast, were isolated similarities in that they were not part of a larger pattern of similarity. Analogies didn't form the circles, but linked them together at various points (Stevens 1994, 186–187). Having sutured wings, for instance, constituted an analogy between Orthoptera, which contained cockroaches and crickets, and Neuroptera, which contained termites and scorpion flies. When analogies between two forms become numerous enough though, they become affinities (Ospovat 1981, 105). In this reliance on affinities, MacLeay was seemingly echoing the views of Aristotle and Buffon that many traits are needed to classify, rather than just the singular or few traits used by Linnaeus in his classifications.

Both the Cuverian system, based on its four embranchments, and the Quinarian system, based on its hierarchical series of five circles, attracted followers. The Quinarian system in particular was debated and adopted by many in England. The Linnean Society, for instance, spent much time discussing it between 1825 and 1840 (McOuat 1996). By the mid-1840s, however, many British naturalists were coming to reject Quinarianism. A young Charles Darwin was familiar with this system and seemed to just assume it in many of his earliest writings, perhaps because it was largely accepted by many of his fellow naturalists in England (Ospovat 1981, 101–103; Hull 1990, 94–95).

The Darwinian Revolution

Darwin, born in 1809, a full century after the birth of Linnaeus, isn't typically known as a systematist, although he studied marine invertebrates as a student at the University of Edinburgh, and in 1846 began an eight-year long taxonomic study of barnacles. He published his results from the barnacle studies in 1851 and 1854 in the two-volume *Living Cirripeda.* For this work Darwin received a medal from the Royal Society of London. This enhanced his status among naturalists, proving that he could do careful, systematic observational work (Stott 2003, 238).

Barnacles typically cement themselves to surfaces, grow hard shells, and catch food with feathery appendages. They are most notable – and problematic – for their conflicting similarities to two very different groups. In the larval stage barnacles, with their feather appendages, seem most similar to crustaceans. In the adult stage they seem most similar to molluscs, with their hard shells cemented to rocks and other hard objects. So should barnacles be classified with the crustaceans in Articulata based on their appendages and larval appearance? Or should they be classified, following Linnaeus and Cuvier, in Mollusca on the basis of their hard shells and rooted lifestyle? This problem of cross-classification generated debate among naturalists of the time (Winsor 1969; Stott 2003, 53). The study of barnacles was significant to Darwin in part because it revealed the great variability in nature. What Darwin found was not a series of discrete, identical morphologies with obvious gaps. Rather there were many gradations among the barnacle forms, and many traits that seemed to cross classify. To classify barnacles, Darwin needed to decide which traits to use, and why. To do so, he would turn to his evolutionary framework. On the basis of this framework, and these years of detailed observations, he came to the conclusion that barnacles were crustaceans rather than molluscs.

Darwin began developing his evolutionary framework in 1837, a decade before his barnacle studies. As is well known, he first published his views in his *On the Origin of Species* in 1859. There were three main claims in the *Origin* that are relevant to his method of classification. First is the transformation hypothesis that new species form and change over time. Second is the hypothesis that natural selection has been the primary (but not exclusive) cause of change. Third is the hypothesis that all species are connected by a common ancestry, in one great branching evolutionary tree. These three claims together gave a theoretical foundation for classification, and for guidance for the development of operational principles.

The transmutation hypothesis is perhaps the least revolutionary of the three. After all, Linnaeus himself had theorized that new species and genera form and change over time through hybridization. And there were several well-known transmutation theories that preceded Darwin, including one presented by his grandfather, Erasmus Darwin, in verse form in *Zoonomia* (1794), as well as the much-maligned theories proposed by Jean-Baptiste Lamarck in *Recherches sur l'organisation des corps vivants* (1802) and Robert Chambers in *Vestiges of the Natural History of Creation* (1844). The most obvious implication of the transmutation hypothesis for classification is that genealogy becomes more relevant than mere similarity for determining membership in a biological taxon. If species change over time, then they cannot be associated with some single well-defined unchanging set of essential traits. To determine how an organism is to be classified, we must therefore look to genealogy not similarity: An organism is a member of a species, genus, order, or class if its parents are. This turn to genealogy was hardly revolutionary, though, as we have seen. Many of those who preceded Darwin, from Aristotle to Linnaeus, Buffon, and Kant, had already accepted a genealogical criterion. What was revolutionary was Darwin's development of the theories that natural selection was the main source of change, and that all species were connected through common ancestry. The first of these two claims is critical to the operational basis of Darwin's method and we will look at it more closely later in this chapter. The second claim about common ancestry is fundamental to the theoretical basis of Darwin's approach to classification.

The only diagram in the first edition of Darwin's *Origin of Species* was of a branching evolutionary tree. By this he illustrated the idea that new species form through branching speciation and divergence from common ancestors. This particular historical and genealogical way of thinking about biological taxa served as the foundation for Darwin's thinking about classification. As early as 1843, six years after he started formulating his evolutionary theory, Darwin was arguing for a genealogical approach to classification in a series of letters to the zoologist G. R. Waterhouse (Wilkins 2009, 132). He finally published his ideas on classification in 1859 in Chapter 13 of his *Origin*, where he began with a reference to the tree diagram.

> I request the reader turn to the diagram illustrating the action, as formerly explained, of these several principles; and he will see that the inevitable result is that the modified descendants proceeding from one progenitor become broken up into groups subordinate to groups. (Darwin 1859, 412)

This group-in-group structure becomes the basis for classification.

> So that we here have many species descended from a single progenitor grouped into genera; and the genera are included in, or subordinate to, sub-families, families and orders, all united into one class. (Darwin 1859, 413)

The idea is that divergence from common ancestry produces a branching tree, and we can use the branch on branch structure of that tree to produce a group-in-group classificatory structure. In effect, the higher-level categories – genera, families, orders, and classes – are to be understood as increasingly inclusive branches on this tree.

But the branching of this tree does not by itself give us a classification. Does a particular branch constitute a genus, a family, an order, or a class? This *ranking* question needs to be answered before a classification can be constructed. According to Darwin, ranking should be based on degree of modification.

> I believe that the *arrangement* of the groups within each class, in due subordination and relation to other groups, must be strictly genealogical in order to be natural; but that the *amount* of difference in the several branches or groups, may differ greatly, being due to the different degrees of modification which they have undergone; and this is expressed by the forms being ranked under different genera, families, sections, or orders. (Darwin 1859, 420)

Aves, for instance, may get a higher rank in a classification because of the large amount of divergence and modification among bird species. Darwin's approach to classification is therefore heterogeneous. Grouping is done on the basis of genealogy, and the ranking of the groups is based on degree of divergence.

So for Darwin, instead of representing God's ideas in the creation, or the mere empirical patterns of similarity and difference, classification was to represent descent or genealogy. This was not an entirely new idea though, because as Darwin pointed out, genealogy had already been used as the basis for classification into species by all those who ignored sexual dimorphism and stages of development:

> With species in a stage of nature, every naturalist has in fact brought descent into his classification; for he includes in his lowest grade, or that of a species, the two sexes; and how enormously these sometimes differ in the most important characters, is known to every naturalist: scarcely a single fact can be predicated in common of the males and hermaphrodites of certain cirripedes, when adult, and yet no one dreams of separating them. The naturalist includes as one species the larval stages of the same individual, however much they may differ from each other and from the adult. (Darwin 1859, 424)

Moreover, we already classify genealogically those organisms whose genealogy we already know – domestic varieties:

> In confirmation of this view, let us glance at the classification of varieties, which are believed or known to have descended from one species. These are grouped under species, with sub-varieties under varieties; and with our domestic productions, several other grades of difference are requisite, as we have seen with pigeons. The origin of the existence of groups subordinate to groups, is the same with varieties as with species, namely closeness of descent with varying degrees of modification. Nearly the same rules are followed in classifying varieties as with species. (Darwin 1859, 423)

Finally, many classifications had been genealogical in effect, even though they were not *in theory* genealogical. When Linnaeus grouped organisms on the basis of their sexual organs, and MacLeay used affinities, both were unwittingly using traits that Darwin thought to be good indicators of genealogy (as we shall see shortly).

One of the main advantages of Darwin's theoretical approach is that, unlike previous approaches, it gave operational guidance. Those shared characters or traits that indicate common ancestry, by virtue of inheritance from a common ancestor, should be used to classify. Those that do not indicate common ancestry are irrelevant. To make this distinction between the characters or traits that indicate ancestry from those that do not, Darwin adopted the terms 'homology' and 'analogy,' used by Richard Owen. Owen was influenced by MacLeay's use of 'affinity' and 'analogy,' but while he adopted 'analogy,' he replaced 'affinity' with 'homology,' and developed his own construal of this distinction. In doing so, Owen was responding to a dispute between Cuvier and Etienne Geoffroy Saint-Hilaire. Cuvier and Geoffroy were both comparative anatomists at the Museum National d'Histoire Naturelle in Paris, but they disagreed about the basic rules of anatomy. Cuvier (as you may recall) adopted a functional approach, whereby traits were assumed to be functional and fitted into the functional whole of each individual. According to Cuvier, we should understand characters or traits by how they served the functional whole of individual organisms. Consequently, the laws of anatomy govern the functioning of parts in individuals (Ospovat 1981, 8). Geoffroy, on the other hand, focused on the extensive patterns of morphological similarity among organisms.

He thought the laws of anatomy should instead be understood as relating these similarities, the *unities of plan* or *archetypes* among groups, irrespective of functioning.

Cuvier and Geoffroy both thought their views to be incompatible. As they saw it, one had to commit to either a functional or a morphological stance, but could not adopt both. Owen saw things differently though, believing that both stances were important and compatible. He focused on Geoffroy's unities of plan, described them as archetypes, and tried to place them within Cuvier's classificatory system, arguing that they served as the foundation for the functional structures of the four embranchments. In doing so, Owen was trying to incorporate insights into both functional and structural similarities. He used the term 'homologue' for structural or morphological similarities that did not necessarily function in the same way, and the term 'analogue' for characters that functioned similarly without necessarily having the same structure (Ruse 1999, 116–124). According to Owen, we could legitimately study organisms in terms of their functions – their analogues, or in terms of their structures – their homologues. We can, for instance, examine the forelimbs of humans and the wings of birds morphologically and structurally to see that they exhibit a similar correspondence of parts, or we can look at them functionally and see that they function in different ways.

This distinction between homologies (homologues) and analogies (analogues) did not automatically solve the operational problem. It didn't tell us whether to classify on the basis of homologies or analogies. But Darwin used this distinction to work out an approach that would. In the letter to Huxley quoted previously, he argued that his approach "will clear away an immense amount of rubbish about the value of characters & – will make the difference between analogy & homology clear." Darwin reinterpreted homologies to be structural similarities due to common ancestry, and analogies to be functional similarities due to adaptation by natural selection. The former then, but not the latter, were a good guide to ancestry and genealogy. In his *Origin*, Darwin explicitly dismissed the value of analogies for classification:

> It might have been thought (and it was true in ancient times thought) that those parts of the structure which determined the habits of life, and the general place of each being in the economy of nature, would

> be of very high importance in classification. Nothing can be more false. No one regards the external similarity of a mouse to a shrew, or a dugong to a whale, of a whale to a fish, as of any importance. These resemblances, though so intimately connected with the whole of life of the being, are ranked merely as "adaptive or analogical characters." (Darwin 1859, 414)

Darwin was therefore disagreeing with Cesalpino and Cuvier, who argued that classification should be based on the most functionally significant traits - the traits required for life. But most importantly, he was arguing that only those traits that indicated common ancestry - homologies - were relevant to classification. He was proposing a special similarity method based on the theoretical foundation of classification as the representation of the evolutionary tree.

Character Classification

Darwin's assumption that natural selection was the main cause of change was fundamental to his operational method. We see this in Chapter 13 of his *Origin*, where he discussed classification in some detail, and proposed a general rule of character classification. Darwin thought that if a shared character has obvious functional significance for some specific way of life, it was likely to be the product of natural selection, and therefore a mere analogy. If so, it was irrelevant to a genealogical classification. This rule might be expressed as follows:

> *General Adaptation Rule*: A shared character trait that is likely to be an adaptation to a particular form of life is not likely to be homologous and is therefore irrelevant to classification.

But this rule does not itself give guidance for determining *when* a shared character trait is likely due to adaptation. It merely tells us to ignore it if is due to adaptation. Darwin proposed a series of corollaries designed to tell us how to identify analogies.

First is the *special habits corollary*: "It may even be given as a general rule, that the less any part of organization is concerned with special habits, the more important it becomes for classification" (Darwin 1859, 44). This corollary is based on the idea that parts associated with special habits, such as flying, swimming, and digging, are more likely to be due to adaptation. Birds and butterflies, for instance, both have wings, which function relative to a special habit – flying. *Wings* therefore should not be used to classify birds and butterflies together.

According to a second and related principle, the *reproductive organs corollary*, organs of reproduction are less likely to be special adaptations and are therefore useful for classification. This is because the organs of generation are typically not associated with special habits of life and the associated requirements of survival (Darwin 1859, 414). Notice that Darwin was here endorsing Linnaeus's use of fructification traits in classification, but he did not use the same justification. For Linnaeus, sexual organs were useful in part *because* they were important for life, and in part because they were highly variable. For Darwin they may be important but they were not due to adaptation by natural selection. So, from Darwin's view, even though Linnaeus had the wrong theoretical justification, he was inadvertently getting the genealogy right by using sexual traits.

A third corollary was *constancy of characters*. Darwin thought that systematists had typically used this principle correctly all along.

> If they find a character nearly uniform, and common to a great number of forms and not common to others, they use it as one of high value; if common to some lesser number, they use it as of subordinate value. (Darwin 1859, 418)

The idea here is that characters constant across species may indeed be important to the species in which they are found, but they are not likely to be adaptations to local circumstances and special habits of life. The similar skeletal structures across vertebrates – birds, dugongs, and humans for instance – are not likely to be adaptive responses to special conditions because these organisms occupy very different environments and have very different habits of life.

Darwin also adopted a *rudimentary organs corollary*. Those organs that appear to be atrophied instances of functional characters, but of little current significance, are unlikely to be adaptations. There are therefore of value to classification (Darwin 1859, 416). Darwin thought rudimentary organs to be important enough that he devoted an entire section to them at the end of his chapter on classification. He concluded this section:

> As the presence of rudimentary organs is thus due to the tendency in every part of the organization, which has long existed, to be inherited – we can understand, on the genealogical view of classification, how it is that systematists have found rudimentary parts as useful as, or even sometimes more useful than, parts of high physiological importance. Rudimentary organs may be compared with the letters in a word, still retain in the spelling, but become useless in pronunciation, but which serve as a clue in seeking for its derivation. (Darwin 1859, 455)

Because rudimentary characters are highly unlikely to be current adaptations to special conditions, they are highly likely to be homologies – present due to common ancestry.

Like rudimentary organs, *embryological* characters are unlikely to be adaptations to local conditions. And the earlier they appear in development, the less likely they are to be adaptive responses, because an embryo in the womb, or a fertilized egg, are typically not subjected to the selection pressures of a particular environment.

> The points of structure, in which the embryos of widely different animals of the same class resemble each other, often have no direct relation to their conditions of existence. We cannot, for instance, suppose that in the embryos of the vertebrata the peculiar loop-like course of the arteries near the branchial slits are related to similar conditions, – in the young mammal which is nourished in the womb of its mother, in the egg of the bird which is hatched in a nest, and in the spawn of a frog under water. (Darwin 1859, 440)

Similarities in embryological characters are therefore more likely to be homologies. And a grouping based on these characters is more likely to be genealogical.

What Darwin was doing in using the general adaptation rule was taking what he thought to be true about the operation of natural selection to determine if a shared trait was homologous and therefore indicative of a common ancestry. He was using the theory of evolution to classify characters, which then could be used to group organisms in a genealogical classification. As we shall see in Chapter 4, this use of evolutionary theory to classify characters as homologies or analogies was also distinctive in the work of systematists in the middle of the twentieth century - and a source of great controversy.

A Darwinian Paradox

Darwin's development of an evolutionary framework is often taken to be one of the great revolutions in science, changing biological thinking and practice. As Theodosius Dobzhansky famously claimed, "nothing in biology makes sense, except in the light of evolution" (Dobzhansky 1964, 449). But it isn't obvious how the Darwinian revolution changed biological classification. Before and after Darwin, organisms were typically grouped into species, which were then grouped into the fixed and nested Linnaean hierarchy of genera, orders, classes, and kingdoms. So while biological classification became richer with the description and inclusion of new plants and animals, to the casual observer the basic framework and approach seemed unaffected by this great evolutionary revolution. Ernst Mayr noticed this, and suggested that the Darwinian revolution had little effect on classification: "As far as the methodology of classification is concerned, the Darwinian revolution had only minor impact" (Mayr 1982, 213).

Perhaps Mayr was justified in his dismissal of the significance of Darwinian revolution to classification, but there is something more to be said. A longstanding problem for classification was the failure to provide an adequate answer to the question about *character selection*: Which characters or traits should be used for grouping organisms together into biological taxa? One obvious way to answer this question is by reference to some assumed purpose of classification. For the early medical herbalists, for instance, the purpose of classification was practical, to ensure the correct formulation of medications. If so, then a classification should be based on *whichever* traits would be most helpful in correctly identifying plants. Those characters that are most variable and easily observed would typically be most useful. But from a modern perspective we can see that this rule is problematic. The most variable and easily observed characters may just be sexually dimorphisms or developmental stages. If so, then this rule would place males and females, and individuals of different developmental stages but still of the same species, into different taxa. If species membership is the most important criterion in the formulation of medications, this rule is counterproductive.

For Linnaeus and the early naturalists who thought biological classification served a theological function, representing the ideas in the mind of God at creation, the choice of characters is also problematic. Should we assume that *all* of God's ideas are equally important, and try to classify on the basis of all traits? But if so, what should we do with traits that cross classify? We might (using Aristotle's example) classify Pinna (a group of bivalve molluscs) with plants based on the fact that they seem rooted. Or we might classify them with animals based on the fact that they have similar circulatory and digestive organs. Does the fact that Pinna is rooted better reflect God's ideas than that it has circulatory and digestive organs? Perhaps we can adopt the view implicit in the approaches of Cesalpino and Linnaeus, that the most important traits are the functional traits that make life and reproduction possible. After all, if God wanted his creations to flourish, then surely those traits that made them flourish were especially significant. But this won't solve the problem of character conflict and cross-classification. Both the rooting and the internal organs in Pinna seemed functional. The character selection problem doesn't seem to be solved theologically, by appeal to the thinking of God.

Perhaps the empiricist approach advocated by Buffon and Adanson, that all traits should be used in classification because that best represents what we observe, can avoid the character selection problem. This approach is also vulnerable to cross-classification, as some characters may imply one classification, while others may imply another, but that may be resolved by simply summing characters. The correct classification is the one that reflects the *most* character distributions. This raises difficult questions about how to identify and individuate characters, as we shall see in Chapter 4. We can take any character and subdivide into multiple characters. Or we can combine characters together into a single more comprehensive character. The problem, as we shall also see in Chapter 4 is this: How can a *purely* empiricist approach tell us how to identify, individuate, and count characters?

Can the genealogical approach associated with Kant and Girtanner help us solve the character solution problem? The obvious answer is that a classification should be based just on those characters that indicate genealogy.

Okay, but which characters indicate genealogy? It isn't clear that Kant and Girtanner had any good answers for that question. Darwin, on the other hand, seemed to have an answer. Use those shared characters that are most likely to be homologous. And natural selection can tell us which characters are most likely homologous. The point is this: Before Darwin there was no obvious way to distinguish those traits that should be used for classification from those that should not. The significance of Darwin's evolutionary approach was not just in his evolutionary theoretical framework, but also in his method of character classification and selection based on that theoretical framework.

Lurking here as well is a more fundamental question: Why should a biological classification be nested and hierarchical at all – whether it represents God's ideas, genealogy, or just the observed complexity of nature? Why should a grouping of things into species that are then grouped into genera, orders, classes, and kingdoms be the right way to classify? Is there something about God's ideas, about genealogy, or about the observed complexity of nature that requires a nested hierarchy? There are, as of the early nineteenth century, no obvious justifications for a hierarchical system, even though the classifications were virtually all hierarchical. There is an explanation though, sketched out in Chapter 1. We classify into hierarchies because when we learn general terms in a language, we learn hierarchies. To learn the terms 'tiger' and 'Siberian tiger' we must learn a hierarchy where the more general category *tiger* is subdivided into *Siberian tiger*. But this is only an explanation, not a justification – telling us *why* we have a hierarchical system, but not why we *should* have a hierarchical system. Here as well Darwin gave us an answer. Classification should be hierarchical because it must be to represent the evolutionary tree with its branch-on-branch structure.

So in 1859, when Darwin published his *Origin*, questions about character selection and classificatory structure had no satisfactory answers, or even a way to think about getting answers. *Even if* naturalists could agree on a single purpose for classification, it wasn't clear which traits should be used to classify, and what kind of system was required. This is the significance of the Darwinian revolution for classification. Darwin gave us a theoretical foundation that in turn provided a framework for determining which traits were relevant to classification and explains why a nested hierarchy is appropriate.

Trends and Tensions

There are several important trends in the early modern history of biological classification, beginning with the herbals of the early Renaissance through to the Darwinian revolution. First was the development of a stable, fixed hierarchy of multiple levels, from the lowest *species* level to the highest *kingdom* level in the Linnaean system, or *embranchment* level in the Cuvierian. Before this period the classificatory terms 'eidos' and 'genos,' and their Latin translations 'species' and 'genus,' were each used at multiple levels depending on the specific inquiry. The only constant was that species were always subdivisions of genera. Second was the development of a naming system by Linnaeus, based on the now familiar genus–species binomial. Before Linnaeus, each naturalist might use a different name for a plant or animal, and classify in an idiosyncratic manner. Third was the proposal and development of theoretical frameworks. An early theoretical framework was based on the idea that universals (general terms) represent the ideas in the mind of God that guided creation. Linnaeus and many of the early naturalists adopted this view. But the culmination of this period seems to be in the development of an evolutionary framework by Darwin. This framework not only gave guidance about what biological classification should represent – common ancestry or genealogy – but also about which characters are relevant – those due to common ancestry, homologies.

There were some tensions in this period as well. First was the tension between practice and theory. The medical herbalists were almost entirely concerned with practice: What is the best way to classify plants to ensure the correct formulation of medications? This required a system that was accurate, easy to apply, and facilitated memorization. But as classification became increasingly theoretical with Cesalpino, John Ray, and Linnaeus, there was an increasing tension. Systematists still used traits that were practical – variable and obvious – but it wasn't clear that classifications based on these traits also served theoretical purposes. It wasn't clear, for instance, that the fructification traits used by Linnaeus really served the purpose of representing divine ideas. Nor was it clear that they could serve as satisfactory substitutes for the use of all traits, which was a goal of a natural classification for both Linnaeus and Buffon.

The second tension was between similarity and history. Echoing Aristotle's distinction between the two senses of 'eidos' and 'genos' as logical universals and as enmattered form perpetuated in generation, many of the early modern naturalists were conflicted in thinking about species, genera, and orders in terms of similarities, and as historical entities or lineages. It was apparent from the beginning that species in particular were historical in that they were lineages composed of parents and offspring. It was also apparent that there were *some* similarities among those organisms that interbred and formed reproductive lineages. But these organisms were not identical. It seems that the species level classification was based on similarity, but species taxa seemed to be constituted by historical relations. And for those naturalists, such as Linnaeus and Buffon, who believed the higher-level taxa to also be historical lineages, this tension extended up the classification.

Finally, there was also a general tension among logic, language, and the world. The first and most obvious problem for systematists and naturalists of this period is that there was disagreement about the application of names to biological taxa. The start of a solution to this problem was perhaps in Linnaeus's use of the genus–species binomial, beginning with the tenth edition of his *Systema Naturae*. But even if there were agreement about the use of a binomial, different naturalists could still use different binomials for the same group of organisms. This problem was addressed by Hugh Strickland and a committee he organized in 1842, that also included Charles Darwin and Robert Owen. This committee endorsed the Linnaean binomial nomenclature, but formulated a series of rules designed to eliminate conflict and ensure stability in naming (McOuat 1996; Amundson 2005, 46–47).

But however the naming problems got solved, there was still a tension generated by the fact that the world is a messy and complicated place, and the names of categories don't reflect that fact. In the living world the gaps or joints are not necessarily clean or clear, as one animal form grades into another, and the patterns of similarities cross classify. But language use requires that we ignore at least some of this messiness in how we apply general terms to the world. When we learn the term 'horse,' for instance, we may need to learn to ignore differences in color and size. As Locke argued, we must learn to *abstract* from our observation of the world. Moreover, when we start putting the categories implicit in our language into a logical system, we also must ignore some of the messiness of the world. To think about horses and birds both as vertebrates, we must ignore many of the differences between the abstract ideas of *horse* and *bird* that in turn rely on the bracketing off of the many differences between individual horses and birds. Linnaeus was aware of this tension, but came down on the side of logic and abstraction. Buffon, in contrast, thought it more important to recognize and accommodate all the complexity in nature. As we shall see in Chapter 4, these tensions persist in the more recent history of biological classification.

4 Evolutionary Taxonomy and the Cladistic Challenge

The Modern Synthesis

We finished Chapter 3 with a look at the Darwinian turn to a genealogical or phylogenetic classification. Darwin argued for two main principles: first, species taxa should be grouped together based on genealogy or ancestry; second, groups of species should be ranked according to degree of modification and diversification. He also gave us operational guidance based on his assumption that natural selection was the main cause of change. In the half century after the *Origin*, interest in classification declined, in part perhaps because a genealogical classification required knowledge of ancestry that may not be available. But focus had also turned to other topics related to the mechanisms of heredity, the cell, and biochemistry, and within the developing fields of embryology, cytology, genetics, behavioral biology, and ecology. These newly developing fields were modeled on the physical sciences, aspired to be experimental, and were becoming less based on zoology and botany. There were some theoretical discussions about the nature of species (Wilkins 2009; Richards 2010), but for many biologists classification seemed more like stamp collecting than science. Ernst Mayr was critical of this attitude in 1942:

> The rise of genetics during the first thirty years of this century had a rather unfortunate effect on the prestige of systematics. The spectacular success of experimental work in unraveling the principles of inheritance and the obvious applicability of these results in explaining evolution have tended to push systematics into the background. There was a tendency among laboratory workers to think rather contemptuously of the museum man, who spent his time counting hairs or drawing bristles, and whose final aim seemed to be merely the correct naming of his specimens. (Mayr 1942, 23; See also Winsor 2001, 11)

Moreover, the adaptationist thinking of Darwin in character classification, based on the operation of natural selection, was no longer in fashion. For many theorists from the late nineteenth century to the first few decades of the twentieth, natural selection was simply not that important to understanding biological processes.

But by the 1930s and 1940s, there were efforts to bring these different biological subfields together, both in substance and practice. This effort, dubbed "The Modern Synthesis," after Julian Huxley's 1942 book of that name, was a merging of both theory and disciplines. We might find the beginnings of this synthesis in the 1930s with the development of theoretical population genetics by Ronald A. Fischer (1890–1962), J. B. S. Haldane (1892–1964), and Sewell Wright (1889–1988). These three used statistical methods first developed in physics to understand population-level processes. They were followed by a group of biologists who incorporated this population thinking into other biological disciplines: Julian Huxley (1887–1975), Theodosius Dobzhansky (1900–1975), George Gaylord Simpson (1902–1984), and Ernst Mayr (1904–2005) (Depew and Weber 1995).

In 1936, Huxley helped form the Association for the Study of Systematics in Relation to General Biology. This led to a session in a 1939 conference on "Genetics in Relation to Evolution and Systematics," that in turn led to an edited volume, *The New Systematics*. In the introduction to this book, Huxley expressed a desire for a synthesis between taxonomy and other biological fields:

> The rise of other branches of biology has been exerting a profound effect on the outlook of taxonomists. Genetics, cytology, ecology, selection theory, field natural history, palaeontology, even developmental physiology – they are all proving to be relevant to this or that aspect of taxonomy. (Huxley 1940, 1)

Although there was little theoretical development of the new approach to classification in this book, the chapters clearly reflected this desire for a synthesis, with topics on classification and genetics, geographic differentiation, ecology, embryology, paleontology, and even philosophy. The theoretical development of this new systematics, which came to be known as "evolutionary taxonomy" or "evolutionary systematics," is more clearly found first in Mayr's *Systematics and the Origin of Species* (1942), and then in greater detail in George Gaylord Simpson's *Principle of Animal Taxonomy* (1961), and the *Principles of Systematic Zoology* (1969) by Mayr and Peter Ashlock. In his *Systematics*, Ernst Mayr contrasted the "old" systematics with the "new," on the grounds that the old way focused on species and morphology, while the new way focused on variation in populations and takes ecology, geography and genetics into account. According to Mayr, this implied that the new systematist is more a biologist than the old:

> The new systematist tends to approach his material more as a biologist, and less as a museum cataloguer. He shows a deeper interest in the formulations of generalizations, he attempts to synthesize and to consider the describing and naming of a species only as a preliminary step of a far-reaching investigation. (Mayr 1942, 7)

As we shall see in this chapter, and again in Chapter 8, Mayr was prescient about the increasing influence of biological theorizing in systematics.

Grouping and Ranking in Evolutionary Taxonomy

As conceived by the evolutionary taxonomists, there are really two domains in biological classification, each with its own distinctive problems and approaches. *Microtaxonomy* is the grouping of organisms into species, while *macrotaxonomy* is the grouping and ranking of species into higher-level taxa. As we might expect, microtaxonomy depends in part on how species are to be defined. In his contrast of the new systematics with the old, Mayr claimed that one of the differences is the reduced importance of species in the new. This does not seem to bear out in his *Systematics and the Origin of Species* though. Most of the chapters deal with species or speciation: allopatric (different place) speciation versus sympatric (same place); animal versus plant species; polytypic species, sibling or cryptic species, multiple species concepts; and species definitions versus species criteria. And in his discussion of the species problem – the use of multiple inconsistent species concepts – Mayr identified and distinguished five conflicting species concepts in use: the *practical species concept*, *morphological species concept*, the *genetic species concept*, the *sterility species concept*, and the *biological species concept*. (Mayr 1942, 115–119) Here he advocated the turn to the *biological species concept*, based on the idea that species are products of processes that produce reproductive isolation.

Reproductive isolation was important, according to Mayr, because it prevents further interbreeding and blending between different populations and makes possible the divergent evolutionary change that produces biodiversity (Mayr 1942, 119–121). According to Mayr, the *definition* of species should therefore be in terms of this process of reproductive isolation. But this definition could not generally be used as a grouping criterion because it could be observed and tested only in sympatric populations – populations that overlapped in space. But because speciation typically came with obvious changes, degree of morphological difference could generally be used for grouping (Mayr 1942, 121). This was not a purely morphological grouping criterion for species though. According to Mayr, "it is always subordinated in importance to biological factors." What he seemed to mean by this is that morphological analysis was always undertaken in light of knowledge of the biological processes governing speciation – geographic isolation, the operation of natural selection, and factors governing fertility and mate preference. Reproductive characters, for instance, may have special significance because of their roles in reproduction, mate preference, and fertility as isolating mechanisms. A general rule seemed to be that those characters that are associated with, or lead to reproductive isolation are of more significance in species groupings (Mayr 1942, 247–274).

The paleontologist George Gaylord Simpson agreed with Mayr about the significance of interbreeding and reproductive isolation, but thought that Mayr's biological concept did not reflect the fact that species evolved over time. His "evolutionary" definition corrected this, by including a temporal element. Species are not just populations at any single time, but population *lineages* that change over time: "An evolutionary species is a lineage (an ancestral-descendant sequence of populations) evolving separately from others and with its own unitary evolutionary role and tendencies" (Simpson 1961, 153). For Simpson, as with Mayr, morphological similarities and differences are important in identifying and distinguishing members of different species, even if species taxa aren't defined by their similarities. After all, ancestral-descendant sequences of populations will change, but only gradually. So a descendant population will share many traits with an ancestor population. But for Simpson's conception of species, there are also special problems related to the distinct *evolutionary roles* of species, roles associated with distinct niches or ways of life in an environment. Here what is known about the operation of natural selection, ethology, ecology, and more will all be relevant. Characters that are indicative of distinct evolutionary roles would be of greater significance and relevance to species groupings. For example, each species of finches on the Galapagos Islands has a distinctive beak shape and size, depending on what kinds of seeds are eaten. For Simpson, as with Mayr, morphological analysis was to be informed by what was known about biological processes.

In *macrotaxonomy* species are grouped into higher-level taxa – genera, families, orders, and classes and so on. Mayr and Simpson adopted Darwin's theoretical approach to classification. Biological taxa are genealogical in that species are grouped together based on "propinquity of descent" or common ancestry. In a term used by Mayr, higher-level taxa are to be 'monophyletic,' containing only the descendant species of a single ancestral species (Mayr 1942, 280). If so, a genus or family must contain only those species descended from a single, relatively recent ancestral species. Significantly, for Mayr a *monophyletic* grouping must contain *only* the descendants of a single species, but need not contain *all* the descendants of that species. So a genus need not contain all the descendants of a single ancestral species. In this way classification, according to Mayr, was only "based on" phylogeny, but did not "express" it.

In his *Origin* Darwin had cautioned that genealogy does not by itself determine a classification. The problem is that there are multiple, inconsistent ways to group species into higher taxa on the basis of common ancestry. Mayr illustrated the problem through a branching diagram representing a hypothetical phylogenetic tree (Figure 4.1).

On this tree there are two main branches, A and B. A splits into four "twigs," 1, 2, 3, and 4. Twig 2 subdivides again into six "twiglets" or species. Twig 4 subdivides into four species. How should we divide these species into higher categories? One could be a "lumper" and group the twelve species of twigs 1–4 together into a single genus. Branch A would then constitute a single genus. Or, one could be a "splitter" and group each of the twigs 1–4 into a single genus. Branch A would then constitute four genera. Notice that each of these competing genus groupings is monophyletic in that they contain *only* descendants of a single ancestral species, so

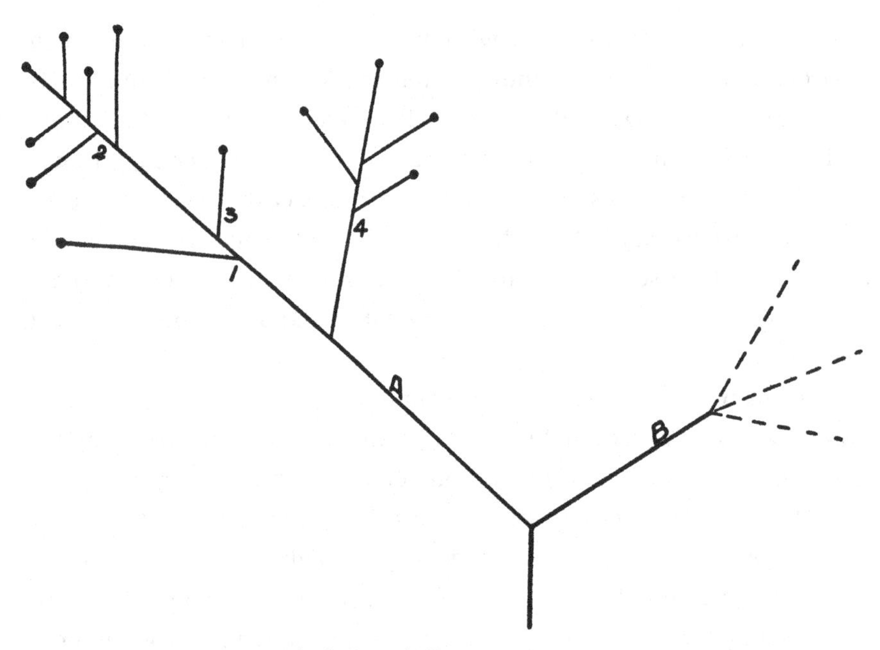

Figure 4.1. Genera on a Phylogenetic Tree.
(From *Systematics and the Origin of Species*, by E. Mayr, 1942, Columbia University Press.)

monophyly by itself cannot decide. How then should we decide the best way to classify these species into genera? Mayr suggested that it is merely a matter of individual taste and convenience:

> The only difference between the lumper and the splitter would be that the lumper would consider the 12 species of branch A convenient unit which is widely separated from branch B and is not so unwieldy as to require subdivision; while the splitter would consider the little gaps which admittedly exist between the species groups (twigs) 1, 2, 3, and 4 to be of sufficient importance to be expressed by 4 different genera, 2 of them monotypic. (Mayr 1942, 182)

This makes the grouping of species into genera subjective, as Mayr admitted:

> The best definition of a genus seems to be one based on the honest admission of the subjective nature of this unit, and it may not be possible to say much more than that "the genus, to be a convenient category in taxonomy, must in general be neither too large nor too small." (Mayr 1942, 283)

Mayr then offered a definition: "A genus is a systematic unit including one species or a group of species of presumably common phylogenetic origin, separated by a decided gap from other similar groups" (Mayr 1942, 283). Genera are to be demarcated by gaps, but when is a gap large enough? Some groups of species are highly variable and have big gaps between each species. In these groups we might have many *monotypic* genera - genera comprised of only a single species. In other groups of species with little variability, the gaps may all be small. Here we might have genera with many hundreds or even thousands of species. Genera groupings in both these cases would be inconvenient, according to Mayr. In the first case of monotypic genera, the genera would be "completely synonymous with the species and would lose its function as category of convenience" (Mayr 1942, 284). In effect the genus groupings would do no classificatory work. In the second case, according to Mayr, "genera with 500, 1,000, or even 2,000 species are very inconvenient, and any excuse for breaking them up into smaller units should be good enough" (Mayr 1942, 283). Such large genera don't serve the function of breaking the groups of species up into manageable, recognizable units.

Mayr allowed some exceptions to this convenience rule that genera should be neither too small nor too big. Sometimes a monotypic genus is appropriate. In cases of low speciation rates but high rates of divergent change, there will be a large gap that should *perhaps* be reflected in classification. Similarly some large genera will sometimes be justified in cases in which there is much speciation but little morphological divergence. In both of these cases, classification reflects both evolutionary processes of diversification and divergence. But according to Mayr, there is no algorithm for deciding when a gap between species groups, or the number of species in a group is large enough, to justify a genus grouping. "It is obvious from this discussion that the delimitation of the genus is, to a considerable extent, a matter of judgment, and that this judgment in turn depends on wide experience and on some intangibles" (Mayr 1942, 286). So given the different experience and judgment of taxonomists, the genera of one systematist may not be the genera of another. Those with lumper tendencies will have fewer but larger genera. Those with splitter tendencies will have more numerous but smaller genera.

On this approach proposed by Mayr, there are two main factors in determining whether a group of species is to be classified as a genus: degree of divergence – the size of the gaps between species – and diversification – the number of descendant species. The genus category is therefore based both on common descent – *monophyly* in Mayr's sense – and evolutionary change – speciation and divergence. The ideal classification is also convenient in that it produces genus groupings that are neither too large and psychologically intractable, nor too small and with no grouping value. Classifications are therefore objective and natural in the sense that they are based on phylogeny and reflect diversification and divergence. But they are also subjective in that first, they serve psychological functions – the breaking up of groups of species into manageable chunks; second, they vary according to the individual lumper and splitter tendencies of each systematist.

These principles were taken by Mayr to also apply above the genus level (Mayr 1942, 291). The grouping of genera into higher-level taxa (families, orders, classes, etc.) will be based on real gaps, and will reflect the real processes of divergence and diversification, but will also be determined in part by how they organize and divide taxa for convenience.

By this time the classificatory hierarchy had expanded far beyond the original five levels of Linnaeus in 1735. In 1961, George Gaylord Simpson, in his *Principles of Animals Taxonomy*, identified twenty-one levels, listed in indented order from highest to lowest:

Kingdom
 Phylum
  Subphylum
   Superclass
    Class
     Subclass
      Infraclass
       Cohort
        Superorder
         Order
          Suborder
           Infraorder
            Superfamily
             Family
              Subfamily

Tribe
  Subtribe
    Genus
      Subgenus
        Genus
          Species
            Subspecies

Moreover, there was the potential for additional levels. According to Simpson, if all the possible super, sub, and infra levels were included between kingdom and subspecies, there would be thirty-four, a number he thought to be more than ever needed in practice (Simpson 1961, 17). As at the genus level, the ranking principles based on divergent change – the size of the gaps – and diversification – the number of species – could conflict with the convenience criterion, producing taxa with too many or too few species. Monotypic higher-level taxa, for instance, would be possible if the gaps were sufficiently large. This was the problem with aardvarks:

> The order Tubulidentata, family Orycteropodidae, and living genus *Orycteropus* of aardvarks exemplify an almost irreducibly simple hierarchic sequence of higher taxa diagnosed and ranked on the basis of divergence and gaps. There is only one living species, so all the taxa named are monotypic in recent fauna … This extremely undiversified group is nevertheless classified as an order because its known members differ as much from any contemporaneous and possibly related orders as the latter differ among themselves. (Simpson 1961, 205–207)

In this case, the size of the gaps between aardvarks and other related species outweighed the demands of convenience. These gaps also reflect the amount of time the aardvark lineage has been diverging. Sometimes the ranking of a group reflects the amount of time the group has evolved separately, because proportionally more divergence will generally be found between groups that have evolved separately for a greater amount of time. But this is not always the case, as there may be proportionally greater gaps over comparable times between faster evolving groups. So a ranking may sometimes reflect not just the degree of divergence, diversification, but also the evolutionary age of a species or group of species.

Convenience suggests that ranking at these higher levels should, as much as possible, be "spaced" to produce equal steps of inclusiveness. Simpson used the classification of horses and zebras to illustrate this principle:

> If, for example, the horses in the narrowest sense (one living species) were placed in one family and the zebras (three living species) in another, the steps from species to family would be exceedingly small in comparison with those remaining to be placed from family to order. The horses and zebras should, by this canon be at most distinct genera and (in my opinion) preferably subgenera. (Simpson 1961, 198)

If horses and zebras were ranked as families, there would be a huge gap between the species level – one horse and three zebras – and the family level, with no intermediate genus level of grouping. As suggested in the last sentence here, the application of this criterion will also be a matter of judgment.

It is not just the ranking of higher-level taxa that requires judgment, though. As we saw earlier (in Mayr's tree diagram) the grouping of taxa can be more or less inclusive depending on lumper and splitter tendencies. But it can also be done in different ways based on how the monophyly requirement is applied. Mayr and Simpson argued that higher taxa should be monophyletic, in that they contain *only* the descendant species of a single ancestral species. *Monophyly*, in this sense, does not require that *all* offspring species be included. There are many competing and inconsistent ways to divide up the offspring of a single species that can satisfy this construal of monophyly. Mayr didn't address this explicitly in his *Systematics and the Origin of Species*, but Simpson did in his 1961 book *Principles of Animal Taxonomy*. Here he gave a series of hypothetical examples of "semiarbitrary subdivisions" (Figure 4.2).

In the evolutionary trees depicted here, all the higher-level taxa (encircled by the dashed lines) are monophyletic in the sense that they contain *only* the descendant species of some single ancestral species. But the lower taxon in A and the left taxon in B do not contain *all* of the descendant species. They are *paraphyletic*. By contrast, the top taxon in A and the right taxon in B not only include *only* descendant species but also *all* the descendant species. They are, according to Mayr and Simpson, "holophyletic." But notice we could also divide up these holophyletic groups paraphyletically. In the top taxon in A we could group the four far

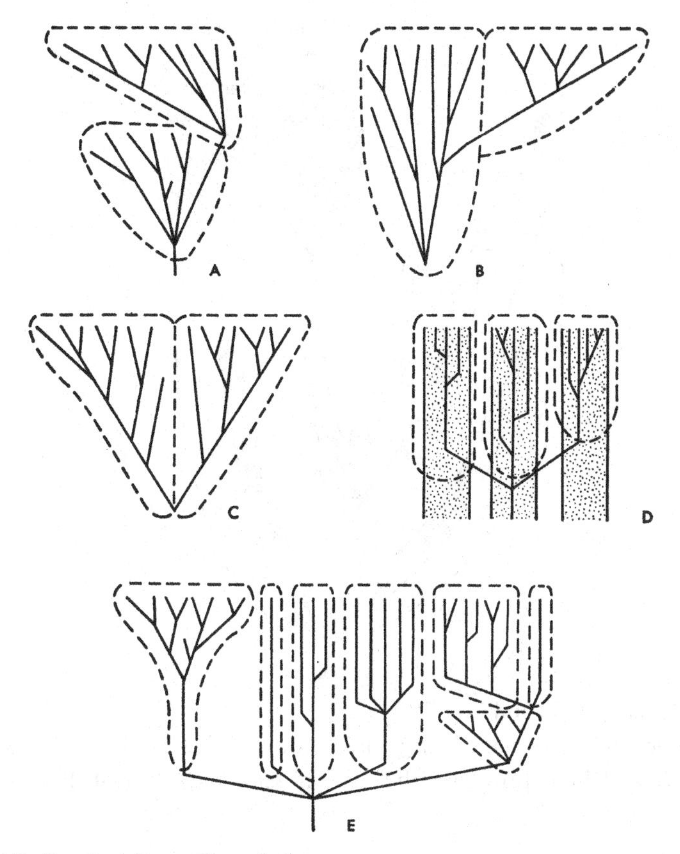

Figure 4.2. Semiarbitrary Paraphyly.
(From *Principles of Animal Taxonomy*. by G. G. Simpson. Copyright ©1990, Columbia University Press. Reprinted with the permission of the publisher.)

left branches together holophyletically, and the four far right branches paraphyletically. But why should we do this? For Mayr and Simpson it would simply be a matter of judgment, perhaps based on degree of diversification and divergence.

One striking and well-known case of *paraphyly* is the traditional classification of reptiles, birds, and mammals. In their 1991 *Principles of Systematic Zoology*, Mayr and Peter Ashlock provide the illustration shown in Figure 4.3.

Here Reptilia contains all crocodiles, squamates, sphenodons, and turtles, as well as the ancestors of birds and mammals – Archosauria, Pelycosauria, and Therapsida. Reptilia is therefore paraphyletic, because it doesn't contain all the descendant bird and mammal species, which are instead in separate taxa – Birds and Mammals. But there are alternative

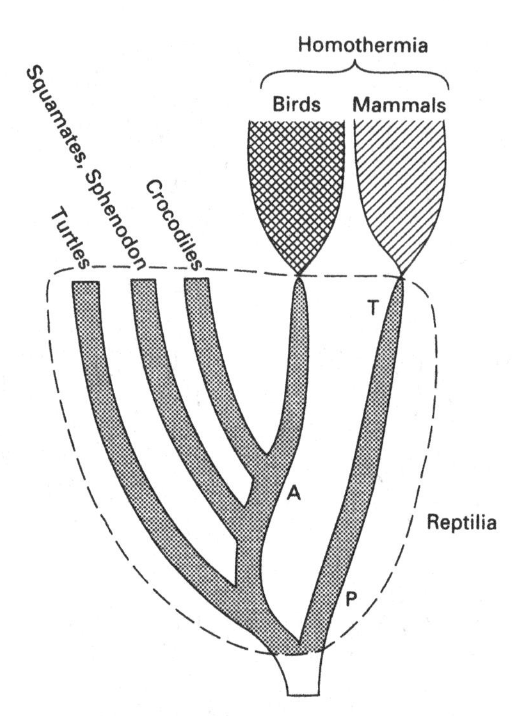

Figure 4.3. Paraphyly and Reptiles.
(From *Principles of Systematic Zoology*, by E. Mayr and P. D. Ashlock. Copyright 1991, McGraw-Hill. Reprinted with the permission of McGraw-Hill Education.)

holophyletic grouping possibilities. First, all the descendant species could be included into Reptilia. In this case, birds and mammals would be reptiles. Birds would also be grouped with crocodiles into a single taxon, and then with crocodiles and squamates and sphenodons into a taxon, and with all three lineages on the left into a single taxon. The relevant question here is: Why group birds and mammals separately from reptile taxa? The answer Mayr and Ashlock give is that birds have acquired a vast new array of characters because they entered a new adaptive zone through a shift to "aerial living," or flight. Here grouping and ranking are related. All else being equal, we might think that Aves and Mammalia should have been given a lower rank than Reptilia, because they are just recent branches in that lineage. But because Aves and Mammalia both have great diversification through speciation, they have great divergent change over time, and there are large gaps between them, they have been given an elevated categorical rank, comparable to Reptilia (Mayr and Ashlock 1991, 249–250).

Character Classification in Evolutionary Taxonomy

There are two main difficulties with evolutionary taxonomy. The first, as we have just seen, is the subjectivity in grouping and ranking. One could classify on the basis of genealogy in multiple ways, with regard to both what group of species gets included in a taxon and what level the taxon will be placed in the hierarchy. So different taxonomists could legitimately classify in different ways based on different inclinations or preferences. But as Mayr and Simpson recognized, there is also a second difficulty. A genealogical classification requires knowledge of phylogeny, which may not be readily available. This may be a temporary difficulty in that although knowledge of phylogeny might not currently be available, it could eventually become available. But this also demands an answer to the question: How is phylogeny determined? Or, in other words, on what grounds should we place each species on the branches of the evolutionary tree, so that we can then group them into higher-level taxa?

Mayr claimed that, as a general rule, those species more closely related evolutionarily will share the most similarities. But he also followed Darwin in rejecting the use of adaptive characters, because similarities here were less likely to be due to common descent:

> Linnaeus and most of his followers for almost a century classified birds by purely adaptational characters. Birds with webbed feet were put into one category, birds with a hooked bill into another, and so on. Eventually it was realized that characters that were adaptations to one specific mode of living were not only subject to rapid changes by selective forces, but also could be acquired in different unrelated lines. Such characters have only a limited value in taxonomic work. (Mayr 1942, 21)

But how can we identify those characters that are relevant? Mayr doesn't answer this question directly, but suggested that the problem can be solved at least sometimes by using many, correlated characters. He believed that the reliance of many characters for classification had in the past, usually but not always, produced good genealogical classifications (Mayr 1942, 277).

Simpson, in his *Principles of Animal Taxonomy*, had much more to say on how to identify those characters that reveal phylogeny. Simpson agreed with Mayr that similarity was in general evidence for genealogy. But he also followed Darwin's example in distinguishing *homology*, "resemblance due to inheritance from a common ancestry," from "resemblance not due to inheritance from a common ancestry," using the term 'homoplasy' rather than 'analogy' (Simpson 1961, 78). *Homoplasy* includes *parallelism*, which is "the development of similar characters separately in two or more lineages of common ancestry and on the basis of, or channeled by, characteristics of that ancestry;" and *convergence*, which is "the development of similar characters separately in two or more lineages without a common ancestry pertinent to the similarity but involving adaptation to similar ecological status." *Mimicry* is also a form of homoplasy, and "occurs when one group of organisms resemble another of different descent within the same community and when the resemblance is adaptively advantageous to the mimicking organism." Finally, there can be *chance* similarity, which is "resemblance in characteristics developed in separate taxa by independent causes" (Simpson 1961, 79).

The important point for Simpson (echoing Darwin) is that only a classification based on homology will be genealogical and based on phylogeny. One based on homoplasy will not. Bird wings and butterfly wings, for instance, are homoplasious, and a classification grouping just birds and butterflies would not be genealogical. So the first and main problem for classification, according to Simpson, is distinguishing homology from homoplasy. And it is largely solved, according to Simpson, based on what we know about evolution: "evolutionary taxonomy must assume as a background the whole body of modern evolutionary theory, much of which is not directly taxonomic in nature but all of which has some bearing on taxonomy" (Simpson 1961, 82). Simpson proposed a series of rules or principles, based on evolutionary assumptions.

The first rule is a *multiplicity rule*, which Simpson claimed was good only for the most part: "Using large numbers of characters probably does tend to reduce the effect of homoplasy on the result, but some effect is always possible and it may be so great as to invalidate the result" (Simpson 1961, 88). The idea here is that similarities based on adaptive response to particular environments (convergence), or chance, are likely to be limited and narrow compared to those similarities based on common descent. So although birds and butterflies are similar in that they both have wings, this is a narrow and limited similarity compared to the many traits where they differ. By contrast, although birds and humans might have striking differences, including the fact that birds have wings and humans do not, there are also many similarities. Not only do birds and humans share an internal bony skeleton, there are many similarities in the structure and relations among their bones. These many similarities indicate common ancestry.

Simpson proposed a second rule, the *minuteness of resemblance rule*. If two organisms are similar in even the most minute details, they are likely to have similar origins. The vertebrate eye, for instance, has its retina behind the optical nerve fibers, while the cephalopod eye has its retina in front of the nerve fibers. So although a human eye and an octopus eye may appear similar in terms of overall shape and design, the details of the structure are different, indicating differences in ancestral states and therefore ancestry. The similarities between the vertebrate and cephalopod eyes are therefore due to independent origins and convergent change.

Closely related to the first two rules is a third, the *complexity of adaptation rule:* "Intricate adaptive complexes are unlikely to arise twice in exactly the same way, hence to be convergent in two occurrences, and the probability of homology is greater the more complicated the adaptation ..." (Simpson 1961, 89). So just as similarities in the minute details in two functional traits are evidence of common origins, so is the overall complexity – the number of functional details shared by two traits. The streamlining in fish and tadpoles, for example, is a relatively simple functional trait, based just on overall body shape. Streamlining could, and apparently did, arise multiple times in different lineages. By contrast, there are many minute details in the anatomy of vertebrate eyes that all function together to produce vision.

The fourth rule is a *special adaptation rule:* "convergence is to be suspected when groups resemble each other only or most closely in some way specifically adaptive to a particular, shared ecology and are otherwise different" (Simpson 1961, 91). By contrast, "characteristics that are more general in nature or that represent a broad improvement in a variety of ecological situations are likely to be more primitive" (Simpson 1961, 100). The basis for this rule, like for the similar general adaptation rule of Darwin, is obvious. Adaptations to some particular environment are likely to be of relatively recent origins and not inherited from relatively distant ancestry. The broad adaptations, functional in many environments, on the other hand, are likely to be adapted not to some single recent environment, but to the many environments encountered by the species of that lineage over long periods of time.

A fifth rule is based on *developmental patterns*. One of the fundamental commitments of evolutionary taxonomy is that classification is of organisms, not parts, and that organisms need to be conceived temporally, as developing over time, from before birth until death. In other words, what is being classified is a *holomorph*, a term introduced by Willi Hennig (to whom we will turn shortly) that refers to "all of the characteristics of the individual throughout its life" (Simpson 1961, 71). Among those relevant characteristics of an individual are the developmental stages and patterns it passes through from conception until death. The idea here is that adult forms are more likely than embryological forms to have changes based on the operation of natural selection, because the embryological forms will have had little exposure to the conditions of life that drive natural selection (Simpson 1961, 89).

Simpson's sixth rule is *convergence of fossil lineages*. He argued that the best evidence for homology often comes from the fossil record - in particular the fossil sequences that get reconstructed.

> Paleontological studies are often based on contemporaneous fossils, and then are just as comparative and based on just as nonhistorical data as those on recent animals. However, when based on sequences in geological time or when relatable to recent faunas, paleontological studies do have a true time dimension and the data are directly historical. In spite of deficiencies in other respects (biased samples, incomplete anatomy, no physiology, etc.) fossils provide the soundest basis for evolutionary classification when data adequate in their own field are at hand. (Simpson 1961, 83)

The idea is that we can infer which similarities or shared traits are homologies if we can reconstruct the evolutionary tree through fossil sequences. Simpson thought that there may be relatively few cases in which the fossil record is complete enough to do so, but when it is, there is direct historical evidence linking traits to common ancestry.

This last rule for determining homology suggests a second task in character classification. If the goal is to reconstruct evolutionary trees, then we need to know more than just which traits are homologous, and due to common ancestry, we also need to know whether a homology is primitive and ancestral - *plesiomorphic* - or specialized and more recent - *apomorphic*. Plesiomorphic traits, because they originated earlier, can be used to group higher-level taxa. The newer apomorphic traits can be used to group only more recently formed, lower-level taxa. If the fossil record was complete in the sense that enough representative individual organisms in a lineage were preserved in the strata to fill out the gaps in a lineage, then this *polarity* could be easily inferred. We could see in the record which shared traits were ancestral and plesiomorphic, and which were more recent and apomorphic. The evolutionary tree could then be easily reconstructed, and the only problem would be how to group and rank the species on the tree. But in 1961 the fossil record had far too many gaps to fully reconstruct the branches of the tree.

Perhaps it would be possible to determine which traits are plesiomorphic and which are apomorphic from larger trends in evolutionary change. If, for instance, evolutionary change is always in the direction of increasing complexity, then the more complex traits would be newer and apomorphic, and the less complex older and plesiomorphic. Simpson considers this possibility:

> Perhaps the most common assumption about evolutionary sequences is that they tend to proceed from the simple to the complex. Such an over-all trend has certainly characterized the progression of evolution as a whole. An amoeba, tremendously complex as it is, is obviously simpler than an ostrich and certainly more closely resembles their common ancestry. Within particular groups of animals, however, this criterion is rarely of much real use. (Simpson 1961, 97)

Simpson's skepticism was based on two problems. First, it isn't always clear how to judge simplicity and complexity. Is an ostrich simpler than a kangaroo? Second, evolutionary change often goes from more to less complex, especially among some parasites. Simpson also considered, and on the basis of the exceptions, rejected the rule that evolutionary change is toward increasing size, and smaller is always ancestral (Simpson 1961, 97–98).

There are some cases, though, where Simpson thought the polarity of a sequence could be determined with some confidence. Certain chromosomal mutations must necessarily occur before others. And because of how duplication and fragmentation works, high chromosome numbers are more likely to be derived from lower (Simpson 1961, 96–97). And some sequences have become so well established in the fossil record that

Simpson claimed they could be adopted tentatively. For example, there seems to be a *reduction rule* relative to toe number.

> For example, among reptiles and mammals, with no known exceptions, the number of normal toes is either five or has been sequentially reduced from that number, so that it can always be assumed with sufficient probability that a smaller number has been derived from the larger. (Simpson 1961, 97)

Simpson also tentatively endorsed a reduction rule relative to other serial structures, such as teeth and vertebrae. The newer apomorphic state is typically then the reduced number of teeth or vertebrae. So on the evolutionary tree, the reduced number of teeth and vertebrae are higher up in the branches.

There are two complications here, however. First, character classifications can conflict. On one criterion, perhaps the *minuteness of resemblance rule*, a trait in two species may be taken as homologous. But another rule, perhaps the *special adaptation rule*, might imply that the same shared trait is a convergent homoplasy. Similarly there might be conflicting criteria for the polarity of a trait. On the *reduction rule*, we might infer that *three toes* is a derived state, but the fossil record might imply that *three toes* in a particular case is instead the ancestral state. In both of these cases, should one criterion be given more weight than another? Simpson did not give explicit guidance about how to resolve these conflicts, but there are some obvious possibilities. First, some rules seem more reliable. According to Simpson, the fossil record, when complete enough, is more reliable than the rule that ancestral traits are simpler and derived traits are more complex. If so, the implications of the fossil record should be weighed more heavily that considerations of simplicity and complexity. This implies that character classification conflicts can potentially be resolved by background information that results in the differential weighting of similarities. We simply put more weight on those characters that seem most reliable. Second, whether a trait is ancestral or derived depends on context – which taxa are under consideration. The forelimbs of humans and birds are ancestral homologies (plesiomorphic) relative to the lower level taxa that evolved after the evolutionary divergence of mammals and birds. But at a higher level, relative to the evolution and divergence of the reptilian ancestors of birds and mammals, the forelimbs of birds *as wings* are newer, derived traits (apomorphic).

There seem to be three main commitments of evolutionary taxonomy. First is the focus on the evolutionary processes that produce and maintain species, and that result in the formation of higher-level taxa. Second is the commitment to a genealogical or phylogenetic classification. All higher-level biological taxa must be monophyletic in the sense that they contain only (but not necessarily all) the descendant species of an ancestral species. Third is the development of operational, character classification rules for the reconstruction of phylogeny, and the grouping of species into higher-level taxa. These rules were intended first, to distinguish between homologies, which indicate common ancestry, and homoplasies, which do not; second, to determine which shared traits are ancestral and which are new and derived. Most significantly, for the debate that followed, these rules relied extensively on a variety of assumptions about the nature and operation of evolutionary processes. And depending on the assumptions, some traits were taken to be more reliable indicators, and weighted more heavily than others. All of these assumptions would be challenged by the "phenetic" approach.

Phenetics

About the same time Simpson and Mayr were working out the details of their evolutionary taxonomy, a radical empiricist approach was being developed by A. J. Cain, C. D. Michener, and Robert Sokal at the University of Kansas, along with the microbiologist and medical doctor Peter Sneath (Hull 1988, 117–120). This approach, called "numerical systematics" by its advocates but dubbed "phenetics" by Ernst Mayr, was empiricist in the sense that it rejected the evolutionary basis of classification in favor of an approach based solely on observable, measurable similarities and differences. In their 1963 book, *Numerical Systematics*, Sokal and Sneath argued against an evolutionary based classification on the grounds that first, evolutionary relationships were largely unknown; second, methods for determining these relationships were circular, because evolutionary taxonomists inferred evolutionary relationships from assumptions of homology, which they in turn inferred from assumptions about evolutionary relationships (Sokal and Sneath 1963, 7). Sokal and Sneath concluded that the only legitimate way to establish evolutionary relationships is from overall resemblance. By doing so, they were rejecting character classification – the determination of whether a shared character trait is a homology or homoplasy, and whether the inferred homologies are ancestral or derived. Unsurprisingly, they also rejected character weighting, arguing that if we cannot tell on independent grounds whether a trait is homologous and ancestral or derived, there are no legitimate grounds for treating one character as more important for classification than another. To treat one trait as more important than another would be arbitrary and subjective.

But also, according to Sokal and Sneath, even if phylogeny were known, a classification could not simultaneously represent all the important features of evolutionary relationships: the time axis, branching order, phenetic relationship, extinction, degree of perfection, velocity of phenetic change, or abundance within a lineage (Sokal and Sneath 1963, 27–28). Any attempt to represent all of these things is destined for failure. By giving birds and reptiles the same rank to represent diversification, for instance, we cannot also represent the fact that the bird lineage is much younger than the reptile lineage, which would require a different ranking. The solution, they argued, is simple. Abandon the effort to represent evolutionary relationships altogether.

But then what is the purpose of classification? According to Sokal and Sneath its function is epistemic in that it can support and license all sorts of inferences about all members of a particular category, or about the traits of each member of that category (Sokal and Sneath 1963, 18–19). For example, a phenetically based grouping of aphids allows for all sorts of predictions.

> Because of high constancy and mutual intercorrelations of characters, such a grouping will carry a high predictive value. Thus, if we read of a new aphid species we can immediately predict a number of characteristics which this species is expected to possess. Being an aphid, it will with almost complete certainty be a plant feeder, possess a particular type of wing venation, be parthenogenetic in part of its life cycle, produce males by nondisjunction of the sex chromosomes, produce honeydew, secrete wax from cornicles or other glands and so on. Since an aphid is a homopteran, we can forecast with some accuracy the general construction of its mouth parts, the texture of its wings and other homopteran characteristics. (Sokal and Sneath 1963, 170)

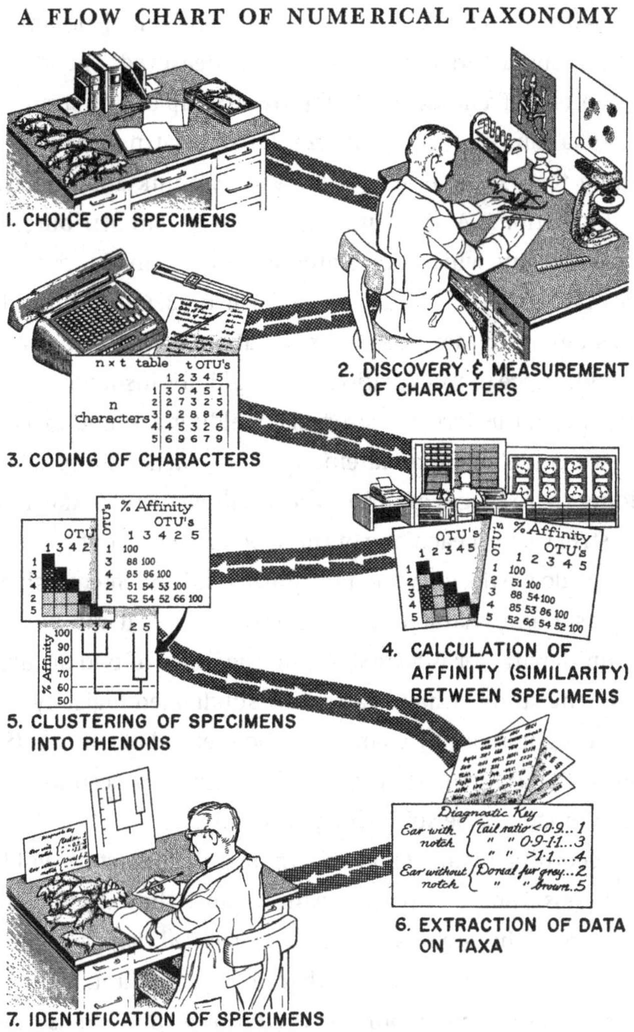

A FLOW CHART OF NUMERICAL TAXONOMY

Figure 4.4. Flow Chart of Numerical Taxonomy.
(From *Principles of Numerical Taxonomy*, by R. R. Sokal and P. H. A. Sneath, 1963, W. H. Freeman and Company.)

This approach seems to combine the requirement we saw in Buffon and Adanson that a natural classification be based on all traits, with the epistemic conception of classification we saw in the approach described by John Stuart Mill, where classifications function primarily to support inferences.

The procedure advocated by Sokal, Sneath and others, is illustrated (rather quaintly) in a cartoon, at the beginning of *Numerical Systematics* (Figure 4.4).

It begins with the choice of specimens – individual organisms to be classified. Then characters are identified and measured. The identification and measurement of characters is far from simple and straightforward though, as Sokal and Sneath recognize. First, what precisely is a character? According to Sokal and Sneath we should think about characters as "units of information." And a "unit character" is "an attribute possessed by an organism about which one statement can be made, thus yielding a single piece of information" (Sokal and Sneath 1963, 63). But this definition won't do. A single statement can be made about a complex of characters, yielding a single piece of information. We could, for instance, make a statement about *three ear ossicles* in mammals, yielding a single piece of information. Or we could make a statement about each – *stapes*, *malleus*, and *incus* – yielding three pieces of information. Or we could make a statement about some structure on each one of these bones, or part of some structure and so on. How do we decide which is right? Sokal and Sneath claimed that what counts as a unit character depends on whether it can be subdivided logically: A unit character is "a taxonomic character of two or more states, which within the study at hand cannot be subdivided logically, except for subdivision brought about by changes in the method of coding" (Sokal and Sneath 1963, 65). Unfortunately, they didn't give an account of when a character can or cannot be subdivided "logically."

But also, according to Sokal and Sneath, not all characters will be appropriate. *Meaningless characters* are irrelevant because they are not a reflection of the genotype of the organism. *Logically correlated characters* are a consequence of other characters, and are therefore redundant. We cannot, for instance, count both hemoglobin and the redness of blood as unit characters. *Partial logical correlations* are irrelevant for similar reasons. Degree of melanization and skin color are partially correlated and therefore cannot both be used. Finally, *invariant characters* are not relevant because they add no information. If all organisms are the same relative to some character, that character does not distinguish any subgroup of the organisms (Sokal and Sneath 1963, 66–68).

On a phenetic comparison of two organisms, there are two fundamental tasks. First, it is necessary to decide whether those organisms share a particular character trait. Second, it is necessary to decide whether the same character in different organisms is in the same or a different state. If so, then the shared character traits are "operational" homologies, in the sense that they are alike (and not in the evolutionary sense that they are inherited from common ancestor). Sokal and Sneath illustrate this by reference to leaf length:

> Leaves on a given series of plants may be long or short. We first have to decide what a leaf is and whether the structures seen on the separate specimens are in fact leaves - that is, are the "same" - or perhaps are other structures such as modified stems. Having decided that they are leaves, we also have to agree on what we shall call a short leaf - perhaps one of less than 3 cm; if so, we shall call a long leaf one that is longer than 3 cm. In this sense all leaves shorter than 3 cm are homologously short leaves, those longer than that are homologously long. The character will now be called "length of leaf," with two states, "short" and "long." (Sokal and Sneath 1963, 70)

Here the leaves were coded into two states - *short* or *long*. A two-state coding could also be in terms of *present* or *absent*. Wings, for instance, could either be *present* or *absent*. Alternatively, coding could be relative to three states as *short*, *medium*, and *long*, or five states as *very short*, *short*, *medium*, *long*, and *very long*.

Once all the characters are identified and coded, then the organisms under consideration are analyzed in terms of these characters and character states to determine the "calculation of affinities" or degrees of similarity. Sokal and Sneath adopted an $n \times t$ matrix approach with n rows representing unit characters and t columns representing fundamental entities being compared. So each entity could be compared in terms of whether it is similar or different from others on the basis of each unit character state. On the basis of these character state distributions, the entities compared would then be grouped into operational taxonomic units (OTUs) based on some algorithm, a "coefficient of similarity." Sokal and Sneath identified three different kinds of coefficients of similarity: association, correlation, and distance. But within each kind of coefficients there were multiple algorithms that group into OTUs differently (Sokal and Sneath 1963, 125–154; see also Panchen 1992, 132–150).

This process produces a hierarchy that can then be represented as a *dendogram*, a branching diagram that indicates similarity groupings by the branches, and degree of similarity by the length of the branches. Sokal and Sneath proposed that the branches on the dendograms be called "phenons," which can then be ranked. Because they thought the Linnaean hierarchy would be insufficiently rich to represent the full range of similarity measures, they proposed that it be replaced with a numerical measure of the degree of similarity based on two ranking principles: first, the internal phenetic diversity of taxa of equal rank should be equal; second, gaps between taxa of equal rank should be equal. (Sokal and Sneath 1963, 203–204). So instead of genera, families, orders, and classes, ranking would be by degrees of similarity: "55-phenons," characterized by branches of 55% similarity, "65 phenons," characterized by 65% similarity and so on.

The phenetic approach never really gained traction, although some of its numerical methods came to be adopted by those with different classificatory goals and procedures. There are several practical reasons why it was never widely adopted. First, a phenetic classification requires a lot of data and consequently an immense amount of effort identifying and coding character traits, not just in newly observed organisms, but also in already well-known and well-studied organisms. If there are little data, classification is simply not possible (Sokal and Sneath 1963, 272). Second, even if the gathering of data on all traits were not so time consuming, the coding and determination of the coefficients of similarity through mathematical algorithms would be daunting for the many systematists not trained in mathematical methods.

But more importantly, there are substantive problems with phenetics. One of the stated aims was to eliminate subjectivity in classification (Sokal and Sneath 1963, 11). But it isn't at all clear that phenetics can do so, as Sokal and Sneath admitted.

> We do not claim that numerical taxonomies are "objective realities"; the fact that a number of slightly different taxonomies may be obtained by different statistical methods is clear evidence that they are not. (Sokal and Sneath 1963, 268)

There are at least two sources of subjectivity in phenetics. The first is in the choice and coding of characters: what gets identified as a character and how it gets coded requires judgment. And it isn't clear that the notions of "unit of information" and "unit character" are helpful in determining what counts as a character. Can either tell us whether *three ear ossicles* is a single or multiple characters? Nor is it clear how to unequivocally code characters. Should a particular character, leaf length perhaps, be coded in two states, three states, or more? Does this all just depend on personal judgment? A second source of subjectivity in phenetics is that different coefficients of similarity generate different OTUs and ranks. There are three kinds of coefficients of similarity - association, correlation, and distance, each with multiple associated algorithms, and it is not clear why on purely *observational* grounds one coefficient of similarity or algorithm is better than another. Some algorithms may be easier to use, but that doesn't seem satisfactory. Perhaps it is up to the judgment of individual systematists. If so, then isn't phenetics subjective, much like evolutionary taxonomy?

There are some deeper philosophical problems here. The first is related to the notions of similarity and difference. The pheneticists saw similarity and difference as purely empirical, based solely on what was observed. We don't seem to need a theory to see that birds are all similar to each other in some ways, and are all different from crocodiles in some ways. We just look and see. If we could base classification on this pure observation, surely we could legitimately test our theories empirically against our classifications. But if classifications depend on some theory, they could not legitimately count as tests of that theory. Sokal and Sneath criticized evolutionary taxonomy because it assumed evolutionary theory and history to identify character traits as homologous, making classification dependent on theory. It therefore could not be an independent source of information for thinking about the assumptions of evolutionary theory. But can there be judgments of similarity and difference that are "pure" - fully independent of any theory? And do we really need these judgments to be independent of theory? We will return to these questions in some detail in Chapter 8.

A second philosophical problem is the relevance of evolutionary history to biological classification. Sokal and Sneath argued that even if we knew phylogeny, we shouldn't use it for classification. For them there was nothing particularly important about evolutionary history for classifying organisms. After all, everything has a history, so biological entities are not unique. And if we don't classify automobiles, nuts, bolts, and screws by their histories, why should we classify organisms by their histories (Sokal and Sneath 1963, 265)? Sokal and Sneath are obviously correct in pointing out that all concrete things have a history. And in some important sense we could think about things like automobiles in evolutionary terms - as descent with modification. While automobiles don't directly reproduce and have offspring, one design may be the basis for other designs.

And designs may diverge and diversify. If so, why should evolution be so uniquely important in the classification of living things? We will return to this question as well in Chapter 8, where we will look more carefully at the relation between evolutionary theory and classification.

Hennigian Cladistics

At around the same time that the details of phenetics were being worked out, an approach dubbed "cladism" by Ernst Mayr, and also known as "cladistics" or "phylogenetic systematics," was being developed by the German entomologist Willi Hennig. His 1950 German book, *Grundzüge einer Theorie der phylogenetischen Systematik*, was read by the German speaking Mayr, and Simpson, Sokal, and Sneath were familiar with its ideas. An extensive revision was translated into English in 1966, titled *Phylogenetic Systematics*. Here Hennig laid out in some detail the basic ideas that would later be adopted by systematists at the American Museum of Natural History in New York, including Gareth Nelson, Colin Patterson, Donn Rosen, Niles Eldredge, Joel Cracraft, Eugene Gaffney, and Ed Wiley (Hull 1988, 142–147). Cladistics would soon spread far beyond the Museum. In 1980, the Willi Hennig Society was formed, and by 1985 it was publishing the journal *Cladistics*, devoted to the application and theoretical development of this approach.

Hennig agreed with the evolutionary taxonomists that classification should be phylogenetic, and based on common ancestry rather than overall similarity. But he disagreed with the evolutionary taxonomists on grouping methods. Mayr and Simpson argued that taxa should be monophyletic in the sense that they should contain *only* species descended from a single ancestral species. Hennig argued that taxa should instead be monophyletic in a more restricted sense, in that they contain *only* and *all* the species that have descended from a single ancestral species (Mayr's "holophyly"). In his Phylogenetic Systematics, Hennig represented monophyly in two ways, in terms of a *cladogram* (II) and in terms of nested set of taxa (I) (Figure 4.5).

There are three things to note here. First, the cladogram looks like the evolutionary trees of Simpson and Mayr, but it represents only relative branching order and monophyletic branches or 'clades,' not degrees of morphological divergence or absolute time. It tells us that there are five higher level monophyletic clades, as indicated in the nested set of taxa in I, numbered 1–5, and that the branching at 1 preceded the branching at 2

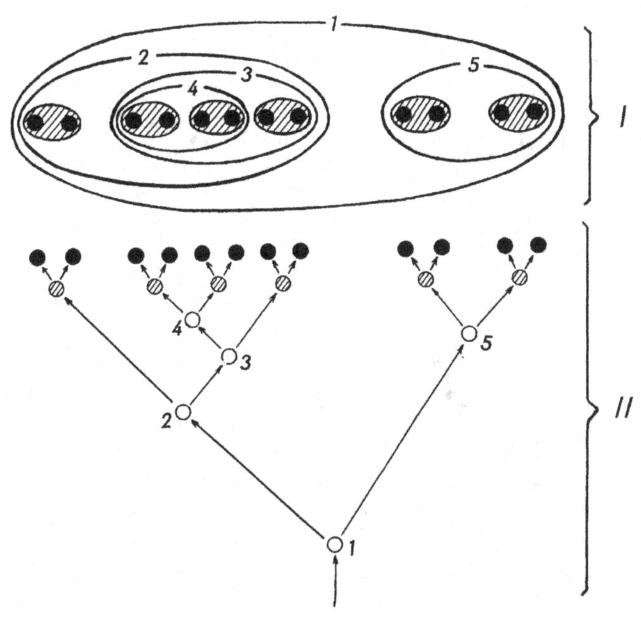

Figure 4.5. Hennig's Monophyly.
(From *Phylogenetic Systematics.* by W. Hennig. Copyright © 1979 Board of Trustees. Used with permission of the University of Illinois Press.)

and 5, the branching at 2 preceded the branching at 3, which in turn preceded the branching in 4. The second thing to note here is that by adopting strict monophyly the grouping is unequivocal. There is only one possible grouping into taxa. Unlike the grouping in evolutionary taxonomy and phenetics, the judgment of different systematists cannot produce different groupings. Third, all of the branching events are dichotomous, in that a single species splits into two different species. This was a methodological convention adopted by Hennig. He did not deny the possibility that more than two species could simultaneously be generated in a radiation by a single speciation event, but thought that the notion of simultaneity was vague enough to allow for the dichotomous representation of multiple speciation events. "On the one hand it is impossible to say exactly when a cleavage process begins and when it ends, and on the other hand indeterminability depends on how long it is before the originally unaltered "daughter species" becomes altered" (Hennig 1966, 212–213). Radiations of multiple species (polytomies) would be used in Hennig's system just to represent cases where methods could not determine a particular dichotomy.

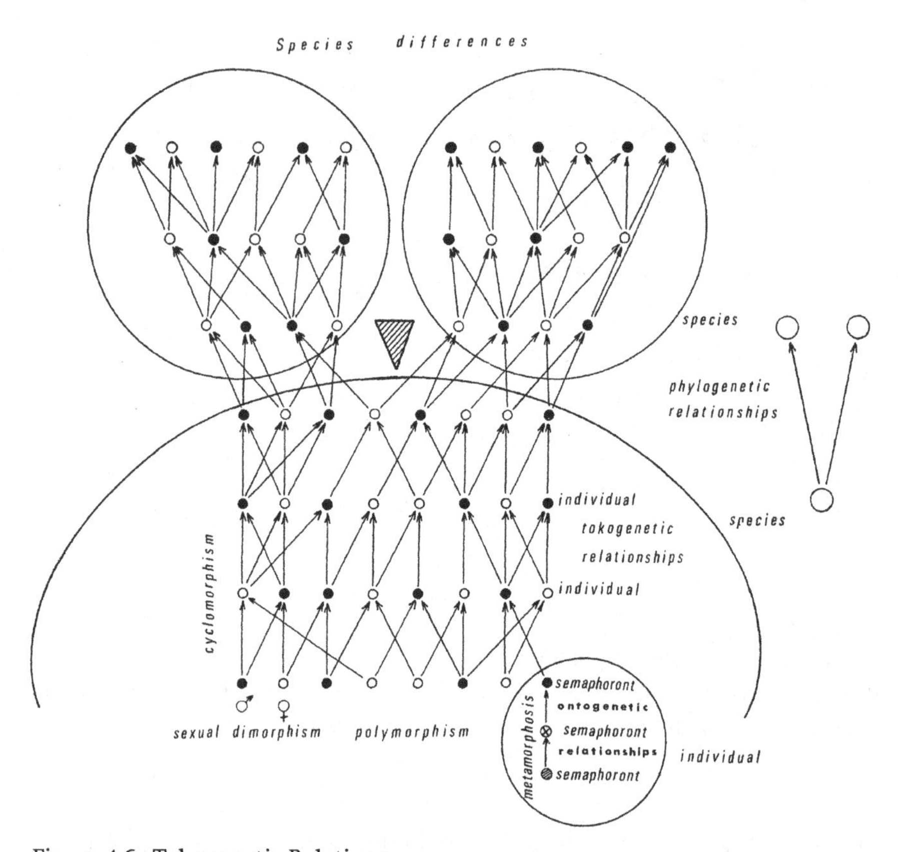

Figure 4.6. Tokogenetic Relations.
(From *Phylogenetic Systematics*. Copyright ©1979 Board of Trustees. Used with permission of the University of Illinois Press.)

Species taxa are represented in this diagram in the black circles at the top of the cladogram in II, and in the nested set of taxa in I. The shaded and unshaded circles are hypothetical stem species that represent the common ancestor of all the species in each strictly monophyletic branch or clade. In Hennig's system, the species category is unique, just as it was in the approach of Mayr and Simpson. Species consist of organisms connected by tokogenetic relationships – genetic relations among individuals that arise through reproduction. Hennig represented these relations in a diagram that shows an ancestral stem species (big semicircle at bottom) and two descendant species (circles at top) (Figure 4.6). The tokogenetic relations are the arrows that connect the small circles that represent individual organisms.

This diagram also shows the branching of a single stem species into two species (with a symbolic wedge), as the small diagram labeled "phylogenetic relationships" to the right indicates.

As the distinction between species and the higher level categories suggests, there are two kinds of ranks, the species rank, which is based on tokogenetic relations, and the higher ranks, which are based on monophyletic relations. The species rank is determined by relations of time, reproduction, and the barriers to reproduction. For the higher-level taxa, there are two main questions. First is the question about *relative ranking*. When are monophyletic clades of the same rank? And when are they of higher or lower ranks? Hennig gave a rule: Sister groups should be given equal ranks. Sister groups are "Species groups that arose from the stem species of a monophyletic group by one and the same splitting process" (Hennig 1966, 139). In Figure 4.5, the two monophyletic clades in taxon 5 each contain two terminal species that comprise a sister group. But the two clades in taxon 2 are also sister groups and have the same rank, even though one clade comprises only two terminal species, while the other clade comprises six terminal species. Acrania - organisms without a skull - and Craniata - those with, illustrate this principle:

> The Acrania and Craniota are certainly sister groups. Consequently they are given the same rank in the system and are (usually) designated subphyla. The Acrania include only about fourteen species, the Craniota tens of thousands distributed over several classes, orders, families, etc. (Hennig 1966, 153)

Ultimately, Acrania and Craniota have the same rank simply because they branched from the same stem species.

The second question is about *absolute ranking*. What rank should a particular monophyletic clade be given? Should it be, for instance, a genus, family, order, or class? In evolutionary taxonomy, absolute ranking, like relative ranking, is determined in part by degree of divergence and diversification. But Hennig proposed the principle that absolute ranking be based on "absolute age of origin," and therefore that "very old groups must always have a high absolute rank" (Hennig 1966, 160–161). So a branch that appeared earlier in evolution would have a higher rank than one that appeared later, and the absolute rank must be in proportion to its age. This is problematic though, in that it would lead to the radical proliferation of ranks. There are many more branches on the evolutionary tree than the twenty or so Linnaean ranks. (We shall return to this problem in Chapter 5.)

While this ranking principle *may* make some sense in classifying living groups, it becomes problematic when classifying extinct species. First, even a single extinct species could be given a relatively high absolute rank, depending on its age of origin – the more ancient the species, the higher its rank. So a single extinct species on a branch would have a much higher rank than a more recent, but much larger and more diverse group of species. A second and related problem is where to place fossils in the classification. Should an extinct fossil species be treated as a terminal species? Or should it be placed into the cladogram at a higher level, perhaps as a stem species? But unless its place in the true phylogenetic tree is known, it isn't possible to know whether any particular fossil species is a stem species for a sister group in a cladogram.

In adopting these ranking principles, Hennig rejected the idea adopted by Mayr and Simpson that degree of diversification and morphological difference should determine absolute ranking. In part this was because of doubts about how to determine degree of morphological divergence. On what grounds, he asked, can we compare the relative degrees of divergence between an earthworm and a lion, and a snail and chimpanzee? This comparison becomes even more problematic when we consider privative traits. Lions and earthworms are alike in that they both lack wings and feathers. Should this be included in an assessment of similarity, difference, and divergence? Hennig concluded that judgments about degree of morphological divergence are ultimately arbitrary (Hennig 1966, 156).

Hennig's approach to classification, based on the branching of evolution and the sister grouping it generates, requires some method of determining branching order and sister grouping. His solution to this problem was based on one simple and striking idea. As new species form and branch off from other species, there will surely be at least one new trait in at least one of the new species: "If a species (reproductive community) is split into two mutually isolated communities of reproduction, there is always a change (transformation) of at least one character of the ancestral species in at least one of the two daughter species" (Hennig 1966, 88). So at each level of bifurcation, there will be a new derived character state associated with at least one of the new species. These derived traits can then be used to identify the species that branched from a single stem species. In effect, we can reconstruct the branching transformation of species by the branching transformation of character traits.

This is an idea we saw less explicitly in evolutionary taxonomy, that some similarities indicate common ancestry, homologies, and that some homologies are more recent and are therefore "derived." The derived homologies indicate more recent common ancestry because they originated in a more recent common ancestor. Hennig used the term 'synapomorphy' for derived homologies, and 'symplesiomorphy' for the older, ancestral homologies. Synapomorphies, but not symplesiomorphies, can be used to identify sister groupings within a more inclusive group of taxa, because they evolved after the origin of the more inclusive group. Symplesiomorphies evolved before that origin so they might be shared by all members of the group. For Hennig, the key to determining monophyletic sister grouping – and therefore classification – is the determination of synapomorphy.

Like the evolutionary taxonomists, Hennig thought we could use what we know (or think we know) about evolutionary processes to reconstruct character transformation sequences. On one principle, based on "ontogenetic character precedence," character transformations in individual development sometimes – but not always – reflect phylogenetic character development. So *sometimes* the characters that develop earlier in individuals are the ancestral, plesiomorphic characters in evolutionary transformation (Hennig 1966, 96). On a second principle, based on the "correlation of series of transformations," if two or more series of transformations are correlated, in that the individual stages of each usually or always appear in different species, the direction of change in one is evidence of polarity in the other (Hennig 1966, 96–101). And paleontology could *sometimes* be useful by indicating which derived traits are ancestral and which are newer and derived (apomorphous): "If in a monophyletic group a particular character condition occurs only in older fossils, and another only in younger fossils, then obviously the former is the plesiomophous and latter the apomorphous condition' " (Hennig 1966, 95). Finally, according to the "chorological method," more closely related species will also be more closely located in space, as the dispersal after speciation takes time (Hennig 1966, 133).

But perhaps most significant for Hennig's method is his "auxiliary principle," which asserts that in cases in which two or more taxa share an "apomorphy," a newer, derived character state, we should *assume* it is an homology and therefore a synapomorphy, unless there is evidence that the shared trait is due to convergence or parallel evolution: "the presence of apomorphous characters in different species is always reason for suspecting kinship [i.e. that the species belong to a monophyletic group]. This is a shift in the burden of proof. Rather than proving homology, we must prove homoplasy" (Hennig 1966, 121–122). In this passage, Hennig justified this shift in the burden of proof by claiming that "phylogenetic systematics would lose all the ground on which it stands" if apomorphies (derived traits) were assumed to be convergences or parallelisms. Unfortunately he does not here explain why. Nor does he explain why we could not adopt a more neutral method, with no a priori assumption of either homology or homoplasy.

Hennig may have followed the evolutionary taxonomists in basing classification on phylogeny, using a special similarity approach based on character classification, and relying on process assumptions to classify characters, but he also rejected those process assumptions used by evolutionary taxonomists based on the operation of natural selection to determine homology and polarity. What is perhaps most significant, though, is that because of the shift in the burden of proof with his auxiliary principle, process assumptions became much less important. With Hennig's auxiliary principle, knowledge about the evolution of traits was unnecessary. We can simply assume shared traits are homologous. This shift in the burden of proof becomes increasingly important in the thinking of Hennig's followers.

Cladistic Clades

The decades following the publication of Hennig's *Phylogenetic Systematics* in 1966 saw an increase in support for the cladistic approach, in terms of both the number and fervor of the systematists adopting it. But there were also many advocates for both evolutionary taxonomy and phenetics. As David Hull described it, the systematists were at "war," battling for control of the systematics journals and conferences (Hull 1988, 158–199). By 1974, cladists had acquired substantial institutional influence. Gareth Nelson was named editor of the leading professional journal *Systematic Zoology*, and then appointed fellow cladist Niles Eldredge a coeditor. Many of the referees appointed by Nelson had sympathies or ties to the cladistics movement (Hull 1988, 339). At the same time cladists also began to take a more prominent role in the meetings of the primary professional organization, the Society of Systematic Zoologists. By 1980 the disputes between the three competing schools – evolutionary taxonomy, phenetics, and cladistics – became so heated that they decided to hold separate meetings: the cladists at the Hennig Society meeting, the pheneticists at the Numerical Taxonomy Conference, and the evolutionary taxonomists at the Society of Systematic Zoology meeting (Hull 1988, 189). Over this same period, many papers were published advocating and developing cladistics methods. And then in 1980 and 1981, three cladistic textbooks were published, the first by Niles Eldredge and Joel Cracraft (1980), and then by Gareth Nelson and Norman Platnick (1981) and Ed Wiley (1981). Over this period there was also a developing rift between two cladistics approaches. One branch, known a "pattern" or "transformed" cladistics, was represented by Nelson and Platnick, while the other branch, "Hennigian" or "phylogenetic" cladistics, was represented by Eldredge, Cracraft, and Wiley.

The phylogenetic cladists adopted most of Hennig's views on both the theoretical basis of classification and the operational framework. They had worries, though, about Hennig's absolute ranking principle that asserts ranks (genus, family, order, class, etc.) should be based on time of origin, where the oldest monophyletic groups or clades would have the highest absolute ranks in the Linnaean hierarchy. One problem, as noted by Eldredge and Cracraft, is that such a system would require knowledge not currently available about the evolutionary tree. Absolute ranking, they concluded, should therefore be abandoned until this knowledge became available (Eldredge and Cracraft 1980, 222–223). A second problem is that new information about phylogeny will often require alterations in the ranks of named taxa. A third problem is the sheer number of ranks required on the Hennigian approach to reflect the differences in relative ranks. The twenty plus ranks of the Linnaean system would be radically inadequate to represent all of the different levels of clades. Eldredge and Cracraft considered some alternatives to the Linnaean hierarchy based on a numerical system advocated by Hennig, a phyletic sequencing system, and an indented-list system. (More on all this in Chapter 5.) After noting the advantages and disadvantages of each, they neither rejected nor endorsed any of these alternatives (Eldredge and Cracraft 1980, 221–233).

The phylogenetic cladists also adopted the operational framework of Hennig – in particular his distinction between synapomorphy and symplesiomorphy. But these cladists were somewhat more skeptical than Hennig about using empirical or theoretical assumptions to determine synapomorphy. For instance, Hennig had argued for the use of the fossil record to determine which traits are derived and which ancestral. But, as Eldredge and Cracraft pointed out, this rule could be misleading in cases where branches with living species evolved more slowly than related branches with fossils at the tips. Fossils might then have *more* derived traits as a consequence of the faster change on their branches, than the more slowly evolving branches with living species (Eldredge and Cracraft 1980, 57). Fossil bird species, for instance, might have more derived traits that extant alligator species. Eldredge and Cracraft were also more skeptical than Hennig was of the value of developmental data, allowing for its use only in cases where it agrees with the primary method based on Hennig's auxiliary principle.

On Hennig's auxiliary principle, if two taxa share a derived state of some character, and if there is no other evidence, that character should be *assumed* an homology and to have a single origin in a common ancestor. On the other hand, if these traits were instead assumed to be homoplasies, and the product of convergence, they would have independent origins in two ancestors. In effect, Hennig's auxiliary principle is a *parsimony principle* that tells us to minimize our assumptions of change by assuming homology where possible. Eldredge and Cracraft explicitly endorsed a parsimony principle for comparing hypotheses (Eldredge and Cracraft 1980, 67), as did Wiley:

> Two or more hypotheses frequently compete against each other in explaining the same data. In such a case, the principle of simplicity (parsimony) is used to pick the hypothesis that explains the data in the most economical manner ... For our purposes in phylogenetics the most parsimonious or simplest hypothesis is that with the fewest ad hoc statements that explain the full array of available data. And we shall prefer such a hypothesis over others that compete for the very same data. (Wiley 1981, 20)

These phylogenetic cladists departed from Hennig's approach though, by using his auxiliary principle as a *primary* method of phylogenetic inference, not merely as a backup in cases where there is no other evidence.

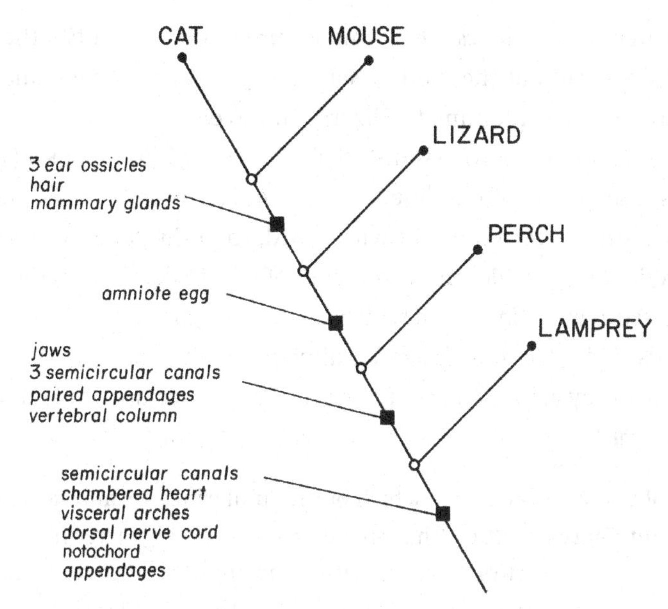

Figure 4.7. Parsimony-based Cladogram.
(From *Phylogenetic Patterns and the Evolutionary Process*, by N. Eldredge and J. Cracraft. Copyright ©1980, Columbia University Press. Reprinted with the permission of the publisher.)

The basic method of cladistic parsimony inference can be applied to any group of three or more taxa, through the successive resolution of sister groupings. Eldredge and Cracraft illustrated the procedure relative to a group of organisms containing *cat*, *mouse*, *lizard*, *perch*, and *lamprey*. The first step is to "undertake a survey of the anatomy of these organisms to note the similarities and differences," and then to group the taxa according to the possession of these similarities. The first grouping will be based on those similarities that all the organisms share. In this group, Eldredge and Cracraft identified the universal similarities: "semicircular canals in the head, a chambered heart, visceral (gill) arches at one or more stages of the life cycle, a dorsal nerve cord, a notochord, and appendages of some kind." Any similarity not found in all five taxa will identify some more restricted subgroup. Amniote egg, for instance, identifies the group composed of lizard, cat and mouse (Eldredge and Cracraft 1980, 23). If we assume homology where possible, in effect minimizing assumptions of change, we can construct a cladogram indicating the phylogenetic relationships of these five taxa, as we can see in Figure 4.7.

In this figure, *3 ear ossicles*, *hair*, and *mammary glands* identify the group containing the cat and the mouse. *Amniote egg* identifies that same group plus a sister group containing the lizard, and so on.

But this cladogram also assumes that the derived homologies (synapomorphies) are correctly identified. If *3 ear ossicles* is instead an ancestral trait, originating before the split with lizard, then the cat–mouse group is paraphyletic, in that it doesn't contain all of the descendant species of the single ancestral stem species in which the trait originated. It wouldn't contain lizards. Eldredge and Cracraft claimed that we can "polarize" traits, determine if they are ancestral or derived, by another use of parsimony in an *outgroup rule*.

> The problem is to determine which of the similarities within the group are evolutionary novelties. This can be accomplished by determining whether the similarities observed within the group are also found in other organisms outside the group. If they are, then those similarities can be postulated as being too widespread to define subgroups, that is, those similarities are not novelties within the set of five organisms. (Eldredge and Cracraft 1980, 26)

The basic idea here is that by examining the most closely related taxa outside the group in question, we can discover which characteristics are new within the group, and which are ancestral. The ancestral traits are those found in outgroups. If *hairless*, for instance is found in the outgroup, then it is ancestral to all the taxa within the group and having hair would then be the derived state. Outgroup analysis can also be applied within the five taxa of the example. The *chambered heart*, *notochord*, and *semicircular ear canal* are found in the outgroup to the group containing the cat, mouse, and lizard, and are by application of the outgroup rule taken to be ancestral and homologous to the ingroup, appearing before any ingroup taxa appeared. This method assumes not just that we can identify outgroups, but also that similarities found in the outgroups are symplesiomorphies – ancestral homologies. The justification for this is parsimony: The best hypothesis is the one that requires the fewest assumptions of change. As homoplasy requires a minimum of two changes, and homology requires only one, similarities with outgroups are assumed when possible to be homologous, and ancestral.

This example provided by Eldredge and Cracraft may be unrealistic in that there are no conflicting character distributions. But often analysis

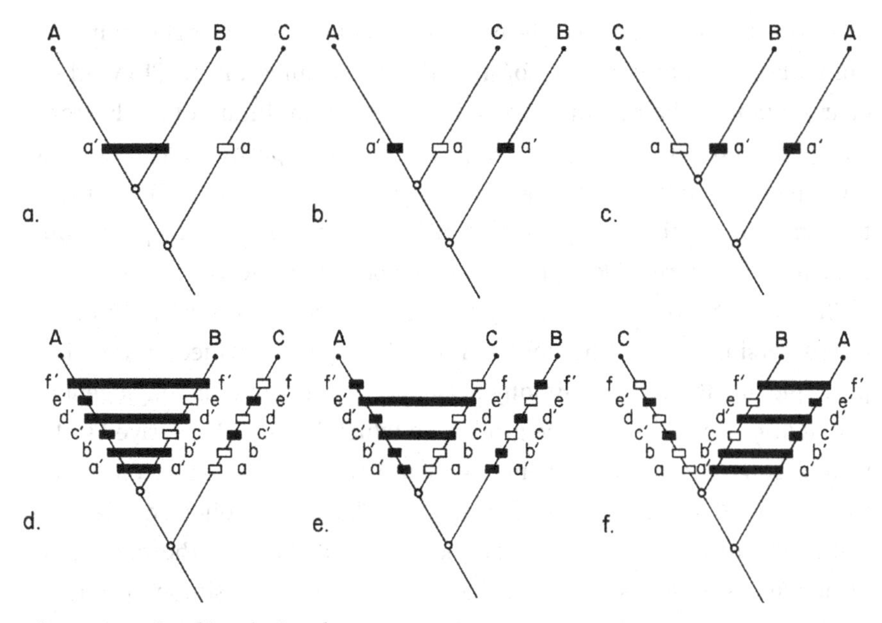

Figure 4.8. Conflicts in Parsimony.
(From *Phylogenetic Patterns and the Evolutionary Process*, by N. Eldredge and J. Cracraft. Copyright ©1980, Columbia University Press. Reprinted with the permission of the publisher.)

produces some characters that indicate one phylogeny, while another set of characters may indicate a conflicting phylogeny. If so, then some shared characters must instead be homoplasies. But which ones? The answer here is also determined by parsimony, as Eldredge and Cracraft illustrated relative to three taxa A, B, and C, and six similarities, a – f (Figure 4.8).

In the first three figures, **a – c**, representing competing phylogenetic hypotheses, a single characteristic 'a' shows how we can use parsimony to determine phylogenetic relationships. In **a**, the hypothesis (AB)C implies a minimum of one change (represented by the solid bar) because a', the derived state, originated once in the ancestor of A and B. In **b** and **c**, there are a minimum of two changes each, one in A and one in B, as homoplasies. So **a** would be the most parsimonious and best phylogenetic hypothesis.

In **d** – **f**, we see how parsimony can resolve conflicting character distributions of derived homologies. In **d**, the phylogenetic hypothesis (AB)C is favored by four derived homologies (long bars) – a' b' d' and f' – implying four changes, but is opposed by two, c' and e,' which must be reinterpreted as homoplasies, implying four more changes for a total of eight.

In **e**, (AC)B, there are two derived homologies (two changes) but four homoplasies (eight changes) for a total of ten changes. In **f**, (CB)A, all six characters must be homoplaseous, indicating a minimum of twelve total changes. So of the three possible phylogenetic hypotheses, (AB)C is the most parsimonious, requiring only eight changes, while (AC)B requires ten, and (CB)A is the least parsimonious, requiring twelve. The parsimony principle therefore picks out (AB)C as the best hypothesis.

But why should we adopt a parsimony principle? Is it because evolution is parsimonious? Is homoplasy rare? Hennig had justified his auxiliary principle, in effect the parsimony principle, by arguing that "phylogenetic systematics would lose all the ground on which it stands" if derived states were assumed to be convergences or parallelisms. But Hennig didn't say *why* his auxiliary principle was necessary. Those who followed offered a justification though. According to Eldredge and Cracraft, the parsimony principle is simply a special application of a general parsimony principle used to evaluate all scientific hypotheses.

> If one views the science of systematics as being subject to the same rules of inference as other branches of hypothetico-deductive science ... science must formulate a criterion by which to judge the relative merits of our close approximations (hypotheses). That criterion, in effect, is parsimony, and it specifies the most preferred hypothesis. (Eldredge and Cracraft 1980, 67)

Wiley gave a similar justification, but with an emphasis on explanation:

> Two or more hypotheses frequently compete against each other in explaining the same data. In such a case, the principle of simplicity (parsimony) is used to pick the hypothesis that explains the data in the most economical manner ... For our purposes in phylogenetics the most parsimonious or simplest hypothesis is that with fewest ad hoc statements that explains the full array of available data. And we shall prefer such a hypothesis over others that compete for the same data. (Wiley 1981, 20)

According to Eldredge, Cracraft, and Wiley, because cladistic parsimony is just the application of a general *methodological* principle of science, it does not require any empirical assumptions about the world, including any about the relative frequency of homoplasies or convergent change. To make this case, cladists sometimes appealed to the views of the Medieval thinker we discussed in Chapter 2, William of Ockham, or the twentieth century philosopher of science, Karl Popper. (See Sober 1988 for philosophical analysis of this debate.)

The Transformation of Cladistics

Hennig had proposed his auxiliary principle to "assume homology," primarily as a supplement to the various assumptions used to classify characters. In the absence of evidence from the fossil record, development, and biogeography, we should assume shared characters are homologous. But for the cladists who followed Hennig, we should *begin* with parsimony, by minimizing assumptions of change. This transformation of Hennig's method made it easier to classify characters - and consequently taxa, in the absence of knowledge of the fossil record, ontogenetic development, or biogeography. It seems that we need not know anything about evolutionary history, or how evolution works to classify! Why not take this a step further and not even assume evolution? A cladistic classification could be treated merely as representing character distributions. This is roughly the view of the "transformed" or "pattern" cladists.

Of the three books published in 1980 and 1981 on cladistics methods, those written by Eldredge and Cracraft and E. O. Wiley were Hennigian or phylogenetic in that that they assumed classification should be based on sister grouping and represent the branching of the evolutionary tree. The third book, by Gareth Nelson and Norman Platnick, *Systematics and Biogeography: Cladistics and Vicariance*, abandoned this assumption. In this book, Nelson and Platnick instead begin with what they call the "problems of form."

> If a biologist discovers a property of an organism, one question immediately raised is "How general is it?" Often the answer is that it has some, but limited, generality: it is true of some, but not all, organisms. Should a second such property be discovered, we can ask not only "How general is it?" but "Is it more or less general than the first property?" If these questions can be answered, sets of organisms can be recognized about which it is possible to make general statements, statements like "All these organisms (1, 2, 3, 4, 5), and no others, have property A" and "All the organisms that have property B form a subset (1, 2, 3) of those that have property A." By this means, sets and subsets of organisms can be defined.

> When these sets and subsets are given names, a classification results. (Nelson and Platnick 1981, 8)

What is striking in this passage is that no reference at all is made to evolutionary history. In doing so, Nelson and Platnick suggest that the purpose of classification is merely to represent order in nature. And cladograms are *just* representations of that order by representing the distribution of general characters. In doing so, Nelson and Platnick distinguish cladograms, which represent "structural elements of knowledge," from phyletic trees, which represent "aspects of evolutionary genealogies" (Nelson and Platnick 1981, 9–14).

> A cladogram is a branching structure joining certain terms (representing taxa) that are related by some unspecified relation. In itself, a cladogram conveys no sense of phylogeny, common ancestry, phenetic resemblance, ecological resemblance, or any other relation that might conceivably join the terms (representing taxa). (Nelson and Platnick 1981, 172)

This isn't to say that evolution is entirely unimportant. Phyletic trees, which represent the branching of evolution, could be inferred from cladograms. The first problem, though, is the fact that a single cladogram is compatible with multiple trees. But which of the compatible trees is the correct one? As Nelson and Platnick saw it, the problem is really to determine which of the branch points in the cladogram are also branching points in the correct phyletic tree. On the cladogram, a branch point is a "synapomorphic resemblance," while on a tree, a branch point is also a speciation event. The cladogram will contain more branch points than the tree, as the species on the cladogram that are ancestors of the other taxa are instead represented in the sister groupings. This is apparent in a figure from Nelson and Platnick, where an assumed phyletic diagram is represented in a cladogram (Figure 4.9).

The trick is to eliminate those branch points on the cladogram that are not speciation events. In part, this is a process of identifying which of the species on the cladogram is an ancestor of the others. In the cladogram here, species 1 is the ancestor of all the other species. The branch on the far left is then one that would be eliminated in a phyletic tree, and species 1 would instead be placed at the trunk of the tree. Nelson and Platnick identify two criteria for determining ancestral species: "an ancestor should be primitive (relative to all supposed descendants) in all known characters; (2) an ancestor should

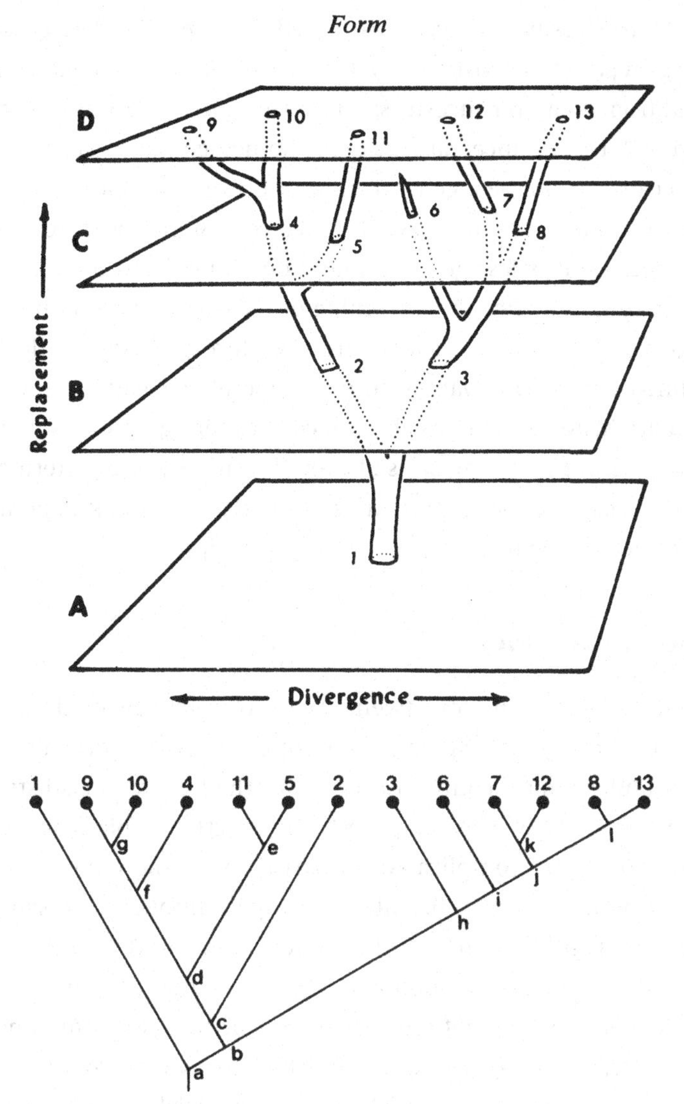

Figure 4.9. Cladogram and Tree.
(From *Systematics and Biogeography*, by G. Nelson and N. Platnick. Copyright@1981, Columbia University Press. Reprinted with the permission of the publisher.)

occur earlier in time than all of its descendants." (Nelson and Platnick 1981, 147). The second criterion may be determined in part by the fossil record.

The operational method of pattern cladistics is much like the method used by phylogenetic cladistics in that it is based on the principle of parsimony to classify characters as homologies and the derived synapomorphies. In pattern cladistics, however, *homologies* are not character traits inherited from common ancestors; rather they are just "units of patterns." In Figure 4.7, for instance, *amniote egg* is homologous for cat, mouse, and lizard, because would be a general trait for this group, not because it was inherited from a common ancestor. And *3 ear ossicles* would be a synapomorphy because it is less general, found only in cat and mouse, and not because it appeared after the branching of the cat–mouse group from the lizard group. The pattern cladists are therefore philosophically similar to the pheneticists who wanted to group merely on pattern and not on evolutionary history. The main difference between them is that while the pheneticists grouped on the basis of overall similarity, the pattern cladists group on the basis of special similarity – nested patterns of synapomorphy, as determined by the use of the parsimony principle.

Philosophical Issues

This brief account of the four twentieth-century schools of classification has neglected many of the complexities of each system, focusing only on the main philosophical commitments. After the two main cladistics positions were laid out in 1980 and 1981, the practice of biological classification became more complicated, in part through the increasing use of numerical methods and of molecular data, especially nucleotide sequences. This trend was apparent in a 1982 conference, the NATO Advanced Study Institute on Numerical Taxonomy, held in Germany. The director of the conference was the pheneticist Robert Sokal, but also attending were some cladists, including Joel Cracraft. That following year the proceedings of the conference were published with Joseph Felsenstein as contributor and editor. In this volume were articles on the use of molecular data as well as statistical methods, and from a variety of perspectives. In his paper, Felsenstein discussed some of the competing parsimony methods, as well as some statistical methods, including Bayesian, and maximum likelihood approaches. The Bayesian approach asserts that the best phylogenetic hypothesis is the one with the highest posterior probability, while the maximum likelihood approach asserts that the best phylogenetic hypothesis is the one that confers the highest probability on the observed data (Felsenstein 1982 and 1983).

By the first decade of the twenty-first century, all three methods of phylogenetic inference had come to be adopted by many phylogenetic cladists. For instance, in a 2011 second edition of a textbook, *Phylogenetics: The Theory of Phylogenetic Systematics*, E. O. Wiley and coauthor Bruce Lieberman include discussion of maximum likelihood and Bayesian methods in addition to the more traditional parsimony methods. Similarly, in his 2012 textbook, *Systematics: A Course of Lectures*, Ward Wheeler covers multiple methods, from distance, to parsimony, maximum likelihood, and Bayesian methods.

This use of non-parsimony-based methods is philosophically significant in part because they tend to involve the use of evolutionary models – assumptions about process – to evaluate phylogenetic hypotheses. The maximum likelihood approach, for instance, relies on models "of change from one character state to another along the branch." Similarly, Bayesian approaches rely on evolutionary models to provide a posterior probability of the tree topology (Wiley and Lieberman 2011, 203–204). This reliance on evolutionary models contrasts with the standard cladist claims of the early 1980s that phylogenetic inference should not rely on the use of *any* process assumptions, at the risk of circularity. But in 2011 Wiley and Lieberman seemed to regard these assumptions as legitimate, and depending on "experience:"

> A phylogenetic researcher might be motivated to incorporate evolutionary models for a number of reasons. For instance, investigators may believe that they understand the evolutionary dynamics of some kinds of data well enough to model their expected evolution over a particular phylogeny. This a priori belief is no different in kind than that of an investigator who weights the performance of certain characters in a parsimony analysis or even the a priori choice of one kind of character over another; any such activities may or may not be warranted. Both investigators are motivated to increase performance of the analysis, increasing what they believe is the veracity of the resulting phylogenetic hypothesis, by applying their experience to the problem. (Wiley and Leiberman 2011, 204)

Philosophically this is more in the spirit of the Hennig of 1966, who allowed for process assumptions, than of the cladistic thinking of 1980 and 1981.

These textbooks reflect what seems to be a developing trend. The basic theoretical framework is phylogenetic and cladistic, in that biological taxa are monophyletic groups, and that classification represents the branching of evolution. But the operational procedures now extend beyond parsimony to Bayesian and maximum likelihood methods, which Wiley and Lieberman claim are really not that different from parsimony:

> The principles used in these different approaches are closely similar. Relationship still means genealogical relationships, synapomorphy is still the mark of common ancestry, and monophyletic groups are the only natural groups regardless of whether one uses a parsimony algorithm or a likelihood algorithm to analyze one's data. That makes us all phylogeneticists, and if you wish to use a label, it makes us all Hennigians. (Wiley and Lieberman 2011, xiii)

Whether or not we are all Hennigians, pattern cladism and evolutionary taxonomy really do seem to be on the decline. This doesn't mean, however, that all the philosophical issues raised in the debates about these approaches have been resolved.

There are three main philosophical disputes in the twentieth century debate over biological classification. First, and perhaps most fundamental, is the debate about the purpose of classification. On one side is the view that a classification should represent evolutionary history. For the evolutionary taxonomists, it should represent genealogy, evolutionary diversification, and morphological divergence. For the Hennigian or phylogenetic cladists, classification should represent only the branching order in evolution and the sister groups produced by this branching. On the other side are the pheneticists, who argued that classification should instead represent overall similarity, and the pattern cladists, who argue that classification should represent character state distributions.

The second philosophical dispute is about the operational procedures used to classify. It is unsurprising that there would be disagreements about how to classify, given the disagreements about the purposes of classification. For the pheneticists, who thought classification should be on overall similarity, this was to be a purely observational procedure. Identify traits in organisms; measure and code those traits; and then, using some algorithm, cluster them into operational taxonomic units. For the evolutionary taxonomists, who wanted to represent phylogeny, diversification, and divergence, classification was to be based on character analysis that uses assumptions about evolutionary processes to reconstruct evolutionary history, diversification, and divergence. For the Hennigian or phylogenetic cladists, who thought classification should represent only the branching and formation of sister groups in evolution, the main tool was the parsimony principle that classified characters on the basis of the number of changes a phylogenetic hypothesis and classification required. Hennig relied on some limited assumptions about process, but initially his followers - in both the phylogenetic and pattern camp, relied almost exclusively on the parsimony principle to identify those traits that indicate sister grouping and phylogeny.

The debate about the use of assumptions to infer evolutionary history is central to a third philosophical dispute, about the relations between pattern and process, on the one hand, and observation and theory on the other. A standard claim of both the pheneticists and the pattern cladists is that pattern is "prior" to process because it is known on a purely observational basis. What we can know best, and most objectively, are the empirical patterns we see. And *only from* these empirical patterns can we infer theoretical process explanations in the form of phylogenetic hypotheses. According to the pheneticists and pattern cladists then, we should avoid the use of theories to classify. This dispute engages some long-standing debates in the philosophy of science about the nature of classification and the relation of theory to observation. This is one of the fundamental tensions in biological classification, and we will look at it more closely in Chapter 8.

5 The Tree of Life

Tree Thinking Before Darwin

In the first edition of Darwin's *Origin* there was only one diagram - a tree diagram, representing the origin of new species through the branching of lineages. As we saw in Chapter 3, this was the basis for his thinking about classification, where the group-in-group structure of the Linnaean system reflected the branch-on-branch structure of the tree. But Darwin wasn't the first to use a tree metaphor in thinking about the relationships of the different kinds of organisms. A century earlier, in 1764, Charles Bonnet had posed a series of tree questions: "Does the scale of nature become branched as it arises? Are the insects and mollusks two parallel and lateral branches of this great trunk? Do the crayfish and the crab likewise branch off from the mollusks?" Two years later the German naturalist Peter Simon Pallas argued that "the system of organic bodies is best of all represented by an image of a tree." And Buffon was using similar language at this time, writing that some groups "appear to form families in which one ordinarily notices a principal and common trunk, from which seem to have issued different stems" (Pietsch 2012, 1–2). In 1801, the French naturalist Augustin Augier published a tree diagram with a trunk, branches, twigs, and leaves. As he explained:

> A figure like a genealogical tree appears to be the most proper to grasp the order and gradation of the series or branches which form classes or families. This figure, which I call a botanical tree, shows the agreements which the different series of plants maintain amongst each other, although detaching themselves from the trunk; just as a genealogical tree shows the order in which the different branches of the same family came from the stem to which they owe their origin. (Pietsch 2012, 1–2; see also Stevens 1983)

Jean-Baptiste de Lamarck (1744–1829) similarly used a tree metaphor a few years later, in 1809, in his *Philosophie zoologique*. Here he drew an upside-down tree diagram "serving to demonstrate the origin of the different animals" (Pietsch 2012, 2, 37).

Darwin's use of the tree metaphor differs from these earlier uses in that the branching of his tree explicitly represented speciation and divergence from common ancestry. Alfred Russel Wallace agreed here with Darwin about evolution and speciation, but in 1855 he also worried about the difficulties of basing classification on this tree:

> If we consider that we have only fragments of this vast system, the stem and main branches being represented by extinct species of which we have no knowledge, while a vast mass of limbs and boughs and minute twigs and scattered leaves is what we have to place in order, and determine the true position each originally occupied with regard to the others, the whole difficulty of the true Natural System of classification becomes apparent to us. (Pietsch 2012, 3)

Wallace's worry about the complexity of the tree seems to contrast with the optimism of Darwin, who in his *Origin* had offered no serious reservations. As we shall see in this chapter, Wallace's worries seem well placed.

This tree metaphor that first appeared in the mid-eighteenth century was perhaps implicit in another way of representing classifications, as bracketed tables. These are tree-like branching tables with the most inclusive groups placed at the left margin, and with brackets indicating progressively less inclusive subgroups to the right. Conrad Gessner was one of the first to use bracketed tables in his *Historia Animalium* of 1555 (Pietsch 2012, 11). Over the next two centuries, this convention became common, and was employed by Linnaeus as well as Lamarck. Although bracketed tables are not explicitly tree diagrams, it is easy to see how their implicit branching structure could be rotated a quarter of a turn counterclockwise into a tree-like configuration with the stem group at the bottom.

Tree thinking wasn't the only graphic approach to classification, though. Map metaphors were also used by Linnaeus, de Jussieu, and de Candolle (Stevens 1994, 164–166). Here the taxa were represented as territories on a geographic map, with the larger taxa occupying more territory than the smaller. The placement of the territories might also represent affinities – the patterns of multiple similarities, so the nearer the territories were, the greater the number of affinities. A map could represent environment and ecological functioning as well. Buffon did so in combining the map metaphor with the tree metaphor in his representation of the races of dogs. He explained:

> In order to give a clearer idea of the dog group, their modifications in different climates, and the mixture of their races, I include a figure, or, if one wishes to term it so, a sort of genealogical tree, wherein one can see at a glance all the varieties. This figure is oriented as a geographical map, and in its construction the relative positions of the climates have been maintained to the extent possible. (Pietsch 2012, 9, 25)

Visually similar to these maps were the geometric diagrams. Best known are the connected circles of affinity in MacLeay's system. But MacLeay also represented analogies as a system of polygons rather than circles. Others have proposed geometric diagrams based on seven circles, a five-pointed star, five five-pointed stars, as well as large circular systems with spirals and vortices (Pietsch 2012).

One other metaphor was more linear. The "scala naturae," based on a metaphor with a ladder, stairway or chain, involved no branching and seemed to represent degrees of perfection. Often God was at the top, exhibiting the most perfection, followed by the seraphim, cherubim, archangels, and angels, to humans, animals, insects, and plants, and finally to the elements – air, water, and earth. This metaphor was explored in Arthur Lovejoy's 1936 book, *The Great Chain of Being*, and taken by many who followed to be about biological classification. Alec Panchen, for instance, focused on how this metaphor was used in the thinking of Charles Bonnet. On Bonnet's scala naturae, humans were at the top of the animal series, followed by monkeys, quadrupeds, birds, fish, reptiles, shellfish, insects, worms, plants, molds, stones, metals, earth, water, air, and ethereal matter. (Panchen 1992, 13). It isn't clear though, that the scala naturae was a classificatory scheme in the same way the tree, map, and geometric representations were. Unlike these approaches, the scala naturae was not hierarchical, and instead seemed to be just a ranking of accepted categories on some single value criterion such as "degree of perfection" (whatever that might mean). In this way the scala naturae is less about biological classification than it is about the value rankings might see with automobiles in terms of build quality, cities in terms of quality of life, or dog breeds as suitability for pets.

Trees after Darwin

The tree metaphor that was used before Darwin, and that was reinterpreted and emphasized by him, became entrenched afterward in thinking about biological classification. We see this in the elaborate and elegant trees of the German biologist, philosopher, physician, and illustrator Ernst Haeckel (1834–1919). While Haeckel had reservations about the central role Darwin gave to natural selection, he quickly became an enthusiastic supporter of Darwinian evolution by divergence from common ancestry. The single tree in Darwin's *Origin* was simple, schematic, and abstract. Darwin did not identify any of the branches or nodes with actual biological taxa. But Haeckel drew trees that did. In his first tree diagram in 1866, "Monophyletischer Stammbaum der Organismen," he identified at the single trunk, the group *monerans*, "simple homogenous, structureless and formless little lumps of mucous or albumionous matter" (Pietsch 2012, 98). There were three great branches of his tree of life, representing the plant, protist, and animal kingdoms. His 1868 book, *Natürliche Schöpfungsgeschichte* (Natural History of Creation), which contained numerous trees, was immensely popular, with thirteen editions over his lifetime. Over the next forty-five years, Haeckel drew hundreds of trees of life with different focal points and stylistic approaches. He drew family trees of animals, plants, mammals, vertebrates, molluscs, jellyfish, and humans. And notably he placed humans on a branch with gorillas, orangutans, chimpanzees and gibbons, and birds on the dinosaur, "Drachen," branch. In one tree he even showed humans, *Homo sapiens*, as direct descendants of *Homo stupidus* (Pietsch 2012, 98–122).

Although there were still some scattered maps, geometric figures, and networks used to represent classification in the last half of the nineteenth century, the trend of the graphic metaphors was clearly toward tree figures. Notable were the attempts to represent evolutionary trees three dimensionally. Heinrich Gustav Adolph Engler, for instance, drew a tree from the usual lateral view, but then drew the same tree as if from above, to show the spread of the branches from the central axis (Pietsch 2012, 131–136). Maximilian Fürbringer drew an evolutionary tree of birds laterally, from two sides, but then also drew cross-sections of the tree at different places. These cross-sections were similar to the map diagrams, with the various cross cuts on the branches represented as circles that indicate the number of member species by the diameter of the circles (Pietsch 2012, 131–141). In these instances, the tree diagrams were intended to represent more than just the genealogy or ancestry of the taxa. The divergence from the central axis typically represented the degree of divergence, while the diameter of the branches represented either the degree of diversification – the number of species in the branch representing each taxon, or abundance – the number of individual organisms.

The beginning of the twentieth century saw two new developments in the tree diagrams. First was the representation of character states on the trees, and then of character trees. Peter Chalmers Mitchell, for instance, drew a tree of the character states of the intestinal tracts of birds. He distinguished between what would later be called by cladists "primitive" and "derived" characters. The "archecentric" characters were ancestral, the "apocentric," relatively new (Pietsch 2012, 149–150, 158). The second development was the inclusion of an absolute time scale on trees that indicated the various geological eras. Oliver Perry Hay, for instance, published a tree in 1908 that showed the phylogeny of living and fossil turtles from the Permian and Triassic to the Pleistocene eras. And in 1917, William King Gregory drew a tree for Henry Fairfield Osborn's book, *The Origin and Evolution of Life*, showing the adaptive radiation of the mammals from the Carboniferous and Permian periods to the Pleistocene. James Small took it one step further and added locations to the branching points, representing the geographic areas where the branches were presumed to occur. And in 1923, Charles Lewis Camp seemed to prefigure Hennig's cladistic approach by recognizing only strictly monophyletic groupings (containing only and all offspring of a single ancestor taxon) based on shared-derived characters, but all represented on a geologic time scale (Pietsch 2012, 149–180). Notable in the 1940s were the spindle and bubble diagrams of Alfred Sherwood Romer, one of the most prolific tree builders. The lineages took on the appearance of bubbles in the most abundant groups, as the lineages expanded, and spindles in the less abundant, as they remained narrow. His diagrams therefore seemed to represent genealogy, geologic time, degree of divergence, and abundance (Pietsch 2012, 181–195).

Tree building continued through the next few decades. William King Gregory (1876–1970), perhaps the most prolific tree builder of all, published trees on winged sharks, invertebrates and vertebrates, reptiles, birds, mammals, birds, bears, dogs, rodents, deep-sea anglerfish, lizards, antelopes, and primates. Many of these trees were of taxa, but many were also of traits. He drew trees, for instance, of primate feet, and hands (Pietsch 2012, 216–241). The trees of this time were highly varied, sometimes using bubble and spindle diagrams, drawings of representative organisms, and circles connected by lines in tinker-toy fashion. But also at this time, we see the first of the tree diagrams based on molecular data. Some of these molecular tree diagrams were produced by the pheneticists, and differed from previous trees in that they no longer represented the branching of evolution, but of overall similarity. Up until this point, the taxa trees had typically been based on character trait trees. Gregory, for instance, seemed to regard them as interchangeable. The hand tree of the primates simply was a taxa tree, with the different hand morphology representing the different taxa, so that the tree simply represented the branching order and speciation in evolution. But with the pheneticists, as we saw in Chapter 4, the tree diagrams represented overall similarity, and explicitly did *not* represent the branching of evolution.

In many ways the trees drawn in the mid to late twentieth century were less interesting than the trees of the late nineteenth and early twentieth century. The trees of Haeckel, for instance, were elaborate and lavishly decorated, while those of Mayr, Simpson, Hennig, Eldredge, and Cracraft were simpler and more abstract. As we saw in Chapter 4, the trees of the evolutionary taxonomists such as Mayr and Simpson were simple line drawings that represented only branching order and degree of divergence. And the cladograms of phylogenetic cladists, such as Hennig, Eldredge, and Cracraft, were further simplified, representing only branching order and sister grouping. The pattern cladists drew what seemed to be identical to the cladograms of the phylogenetic cladists, but interpreted them as mere character trees, representing the distributions of synapomorphies (derived homologies) rather than evolutionary branching events. The most recent trends of twentieth century tree thinking have been toward the more literal interpretation of tree diagrams, as representing the branching of evolution that we saw in the phylogenetic cladists. We see this in the current attempt to reconstruct the single, grand tree of life.

The Tree of Life

In what is now called "phylogenetic biology" or "phylogenetic systematics," the goal is to reconstruct the branching sequences of all the species on Earth. We see this in two recent textbooks: *The Tree of Life: A Phylogenetic Classification*, by Guillaume Lecointre and Herve le Guyader (2006), and *Tree Thinking: An Introduction to Phylogenetic Biology* by David Baum and Stacey Smith (2013). The focus in *The Tree of Life* is on the tree itself and its three main branches, Eubacteria, Archaea, and Eukaryotes, along with methods for constructing the tree based on parsimony and statistical methods. The emphasis in *Tree Thinking*, on the other hand, is not so much on the great tree of life itself, but on the philosophical framework for thinking about phylogenetic trees, with sections on "Interpreting Trees," "Inferring Trees," and "Using Trees."

This commitment to tree thinking is perhaps most striking in the web-based, collaborative tree of life projects. The Tree of Life web (tolweb.org) was first put online in 1994 by David Maddison, who was then joined by others, including Katja-Sabine Schulz and Wayne Maddison. The premises of this project are that there is a single tree of life, and this tree is central to our understanding of life, as the promoters of this project declare:

> We now see all life on Earth as intimately connected in a single tree-like structure of flowing nucleotide sequences, housed in bodies of organisms they help build. This tree is billions of years in age, with the myriad branches and millions of extant leaves. The existence of this tree, and that each of us is part of one of its leaves, is one of the most profound realizations that we have as a species have achieved. (Maddison, Schultz, and Maddison, 2007, 20)

This tree is significant, not just because it represents the branching structure of evolution, but also because of a sense of wonder and understanding it might provide:

> Far from being simply an academic exercise, this approach promotes a deeper appreciation of the diversity and history of life, and it provides for a more profound understanding of biological patterns and processes. We hope it also provides a sense of wonder about the existence of the several-billion-year old tree that has produced all organisms on Earth, including *Homo sapiens*. (Maddison, Schultz and Maddison, 2007, 21)

The tree, as conceived here, is a series of interconnected nodes: the root node that represents the ancestor of all life, branch nodes with descendant species that represent monophyletic clades, and terminal nodes or leaves, which represent species with no descendant species. The branches and leaves have associated web pages that provide more technical information about each clade or species, along with "treehouse" pages for children and the general public. The branch pages give the phylogenetic tree of each group along with an indented list of taxa with Linnaean ranks, and information about subgroups. Leaf pages are usually about species, but might be about subspecies, varieties, and strains. More detailed information about each branch or leaf is also given through links to articles and notes.

The tree diagrams in the Tree of Life web project reflect the views associated with Hennigian or phylogenetic cladism. The branching order of taxa is represented but not the degree of divergence. Like the cladograms of the early phylogenetic cladists, the branch nodes represent hypothetical ancestors, but are not associated with actual fossils. When methods of phylogenetic inference cannot resolve the branching into a dichotomy, polytomies (three or more branches at a node) are represented on the tree. And there are no ranks on this tree of life. No clade is deemed a genus, family, order, or class, although pages associated with branches discuss conventionally accepted Linnaean ranks. A branch page may discuss the various genera or families on that branch, and often give a detailed classification with Linnaean ranks. Cephalopoda, for instance, is shown as a class containing subclasses, divisions, superorders, orders, suborders, superfamilies, and families (http://tolweb.org/Cephalopoda/19386/2012.11.10). In spite of all this, the authors deny that the tree "is about classification." They qualify this, though, by acknowledging that there is *some* relation between classification and phylogeny:

> Of course, there is a general correspondence between classification and our ideas about phylogeny. Most researchers feel that the groups we give names to in our classifications should correspond to single branches or clades on the phylogeny: if we wish to talk about the history of a group, it is easier to do so if the named entities are cohesive portions of that history. (tolweb.org/tre/learn/concepts/classification.html, January 12, 2015)

Unfortunately, the authors on the Tree of Life project do not here say anything more about precisely what the relation is between the tree of life and classification.

Another recent project is the "Open Tree of Life," which was supported by a National Science Foundation grant with eleven principal investigators at ten different institutions. Like the Tree of Life project, it is collaborative. It is, as the organizers describe, a "phylogenetic synthesis," in that the tree of life is based on a "merging" of previously published and expertly "curated" trees of various groups of organisms. As of December 2014, there were 153,109 branches (clades) based on 483 distinct phylogenetic trees. The plans are for a "first draft tree" with 2.3 million "tips" or leaf nodes (Smith, Cranston, and others: http://dx.doi.org/10.1101/012260). The Open Tree of Life, unlike the Tree of Life project, integrates classification with the phylogenetic tree. The project researchers see this as necessary to building a single tree from all the different sources.

> A major, unappreciated challenge to any large phylogenetic analysis involving multiple studies is the need to map organisms to a common taxonomic framework. Tips in each input phylogeny, which may represent different taxonomic levels, must be mapped to known taxonomic entities in order to align phylogenies from different sources. (Smith, Cranston and others: http://dx.doi.org/10.1101/012260)

The organizers also propose a "reference taxonomy," the Open Tree Taxonomy, that synthesizes or combines multiple systems in use for different branches of the tree, such as the Index Fungorum, and contains both named Linnaean groups and unnamed taxa.

The idea behind these collaborative tree of life projects is certainly attractive and in line with the thinking of Darwin 150 years ago in his *Origin*. But the construction of a comprehensive tree of life is not so simple or straightforward. There are typically many conflicting evolutionary trees worked out for each group of organisms and that form the basis for the branches of the tree. One researcher may reconstruct the phylogenetic tree of bats one way, for instance; another may reconstruct it a different way; and a third may reconstruct it in yet a different way. The problem is that different researchers may apply different methods based on parsimony, maximum likelihood, or Bayesian approaches; use different data sets; and arrive at different outcomes. But how can we decide which subtree to incorporate into the one grand tree of life if we aren't sure which is correct? One way to resolve this conflict is through consensus trees. On *strict consensus*, only those branches that are on each of the competing trees get incorporated into the tree of life. But that method typically leaves many of the branches on the tree of life unresolved, represented by polytomies, with three or more branches at a node. If the conflicting trees are very different this might produce twenty or more branches at a single node. This problem is mitigated, but not solved by adopting a weaker, *majority consensus* criterion that simply requires a majority of trees to agree about a branch.

Regardless of the difficulties, advocates of these projects see great value in the reconstruction of the comprehensive tree of life. First, and perhaps most importantly, the tree of life is usually seen as valuable simply for representing the evolutionary history of species. Surely the reconstruction of evolutionary history is valuable for its own sake, much like the efforts of geologists to understand the history of the Earth and cosmologists to understand the history of the universe. Second, with the inclusion of character data, the tree can represent the evolutionary history of characters. We can then better see the patterns of character change. We can see, for instance, when particular characters originated, when they became modified, and perhaps how they became modified. We might also see multiple origins of particular characters, and that some characters are fleeting while others are long lasting. We might see evolutionary associations or dissociations among characters. And perhaps with all this information we can formulate generalizations about character change. Third, we can correlate other information, from biogeography and ecology perhaps, with the tree. We can note how geographic distributions or ecology correlate with character changes, as each branch may have distinctive geographic locations or ecologies. This information could help us think about evolutionary processes related to speciation and adaptation by natural selection. And perhaps we might be able to test process hypotheses on the basis of the tree of life. Fourth, we might use the tree of life for purely pragmatic purposes. Some organisms are the source for medicines, and closely related organisms may also have medicinal value. On the other hand, poisonous or pathogenic organisms might also have similarly poisonous or pathogenic relatives. We can potentially look at the branches of the tree of life to predict the likely traits of organisms on any particular branch of the tree.

There seem to be three general functions of the tree of life. The first is *descriptive*, by representing the hypothetical patterns of evolutionary change in the proliferation of species and the modification of character traits. The second function is *heuristic*, by representing and integrating biological information in such a way that it can be used to generate new hypotheses and ways of thinking about biological processes and entities. Trees are a way of thinking that helps us extend current thinking and develop new ways of thinking within the evolutionary framework. The third function is epistemic, by providing the grounds for generalizations and predictions – thinking about biological information in systematic ways.

Ranking and the Tree of Life

As the brief history of tree thinking at the beginning of this chapter reveals, trees have long been an important way of thinking about biological diversity, extending from pre-Darwinian thinking to the recent predominance of phylogenetic cladistics and the newly sprouting tree of life projects. There is no reason to think that trees and tree thinking are going away anytime soon. But as A. R. Wallace saw, in the passage quoted earlier this chapter, there are significant complications to the use of trees as the basis for biological classification. First is the ranking problem. Even if we can reconstruct the tree with confidence, we still don't have a way for determining how the clades representing the branches should be ranked. Should a particular branch be designated a genus, family, order, or class – or one of the many ranks in between? As we saw in the last chapter, Hennig proposed that we rank according to age of origin – the older the branch, the higher the rank. There is something intuitively right about this proposal. The very earliest branches in the tree of life are the origin of the highest categories of organisms – the kingdoms Eubacteria, Archaea, and Eukaryota. And within each branching structure of the tree, we might plausibly think the earlier branching clades in general should be given higher ranks than the later branching clades. But given the multitude of branches on this tree – more than 150,000 so far in the Open Tree of Life – it isn't possible for each branch to have its own Linnaean rank. Presently, there are only twenty to thirty ranks. And as we come to know more about the phylogeny of life, we will be adding many more branches to the tree, exacerbating this problem.

But if the Linnaean ranking system is inadequate, the Linnaean naming system must also be inadequate. On this system, as we have seen, each species is given a binomial that indicates genus and species membership, as in *Homo sapiens* – the species *sapiens* within the genus *Homo*. Ranking is implicit here, in that the first name indicates the genus rank, and the second the species rank. But the Linnaean system also implicitly references rank in the names of higher taxa. On one standard classification, humans are members of the subtribe Hominina, the tribe Hominini, the subfamily Homininae, the family Hominidae, and the superfamily Hominoidea. The suffixes here indicate rank: 'ina' for subtribe, 'ini' for tribe, 'inae' for subfamily, 'idae' for family, and 'oidea' for superfamily. Among plants, the standard suffixes are 'eae' for tribe, 'oideae' for subfamily, 'aceae' for family, 'ales' for order, 'opsida' for class, and 'phyta' for phylum (Baum and Smith 2012, 130). If we don't have a sufficient number of ranks, we are not likely to have a sufficient number of rank-indicating suffixes.

Moreover, if classification is based on the tree of life, and if names of taxa are tied to ranks, via suffixes or any other method, then naming becomes unstable. As the tree of life is filled out, and modified based on new information, the names of taxa will change. At the lowest level, the binomial may require revision. If, by virtue of a revised phylogeny, a species is found to belong to a different genus, the binomial will have to change because the genus name changes. The bee species *Paracolletes franki*, for instance, is now believed to belong to the genus *Leioproctus* instead of *Paracolletes*. This name must then be changed to *Leioproctus franki*, but there is already a species in that genus of the same name. So an entirely new name for this species is required. Recently, because of revised phylogenies, 288 species of bees were reclassified into different genera, requiring that they all be given a new genus name (Ereshefsky 2001, 224).

This instability is found at higher taxonomic levels as well. This can happen in several different ways. First and most obviously, changes in rank often require a change in name. This can potentially happen through the addition of a new branch to the tree. If the discovery of new species and branch on the human branch, for instance, indicated that the tribe known as Hominini should instead be ranked as a subfamily, then that group must be renamed with the 'inae' suffix. Or if new information suggests a taxon is paraphyletic, and doesn't include all descendant species of single ancestor, then that taxon will be eliminated altogether in favor of a new monophyletic group. Although it is possible the new monophyletic group can retain the name of the paraphyletic group, it is also likely that many of these revisions will require adjustments in naming scheme (Ereshefsky 2001, 224–225). Most famously, this has already happened with Reptilia, the taxon that contained turtles, lizards, snakes, and crocodiles. But because it is now believed that birds and mammals are also descendants of the same common ancestor – birds and mammals are located on the same branch of the tree – they must be included in the monophyletic taxon Amniota that replaces Reptilia. This has also happened on the human branch, where the group of species that includes *Homo sapiens*, but not the chimpanzees and other apes, was once known as Hominidae, while the chimps, gorillas, and orangutans were classified separately in Pongidae. But the now standard view is that chimps and humans have a more recent common ancestor than either had with the gorillas and orangutans. This made Pongidae paraphyletic. The newer monophyletic group has humans alone in Hominina, together with the chimps and bonobos in Hominini, with the chimps, bonobos, and gorillas in Homininae, and with chimps, bonobos, gorillas, and orangutans in Hominidae. In general, with each revision of phylogeny in the tree of life, there will likely be some change in the names of taxa.

The costs of this instability are obvious. If a revision of the tree requires a change in the binomial of a species, then everyone who has ever learned the original name must forget it and learn the new name. Similarly for the higher taxa: For every branch added to the tree, or every taxon eliminated because of paraphyly, a name change is required. If so, then classification and textbooks become outdated. And those who learned the old naming scheme would need to learn the new. Moreover, we would need to keep track of the name changes, so that older references to a taxon can be translated into the new naming scheme. This instability seems inevitable if we combine a tree-based system of monophyletic classification with the Linnaean naming system. No wonder the creators of the Tree of Life web don't want to assign ranks to the branches.

Several solutions have been proposed to this suite of ranking problems. One solution, the *annotated Linnaean classification*, is conservative in the sense that it preserves the basic Linnaean framework, but uses conventions to accommodate all the new levels required for the many branches. On this approach, the current Linnaean names and ranks are to be preserved as

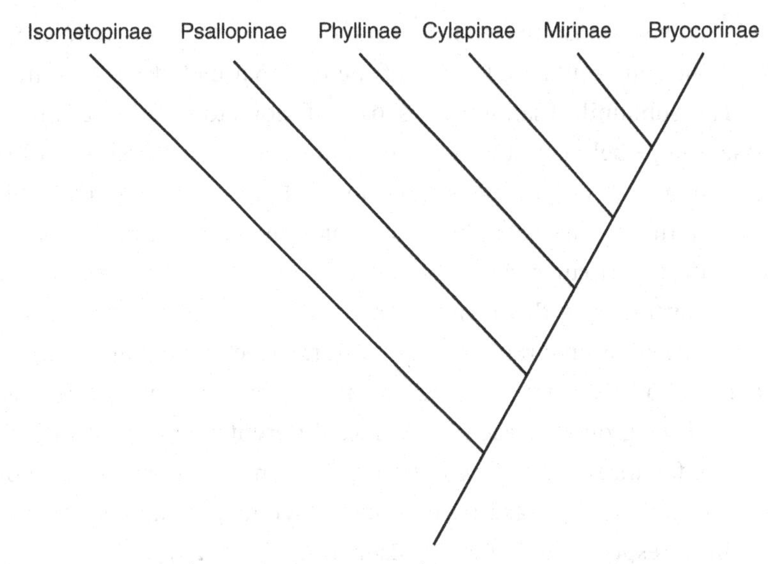

Figure 5.1. Sequencing and an Asymmetric Tree.

much as possible, but with a series of conventions intended to accommodate the new levels of the hierarchy required to represent the branches of the tree of life (see Wiley 1981 and Wiley and Lieberman 2011). A sequencing convention allows the branches of the tree to be represented by the listing sequence of taxa. For instance, the branch of the tree containing the "capsid bugs," which include thousands of species of small leaf bugs, tree bugs, or grass bugs, is classified as the family Miridai. On this branch of the tree there are five sub-branches (see Figure 5.1). If each branch (monophyletic group) were given its own rank, there would be a need for six ranks within the family Miridai.

But by listing the taxa in order of branching sequence, the taxa can all be given the same rank - subfamily - and the classification can still represent branching order and the inclusive hierarchy of monophyletic taxa. That classification could then be represented as below:

Family Miridae

- Subfamily Isometopinae
- Subfamily Psallopinae
- Subfamily Phyllinae
- Subfamily Cylapinae
- Subfamily Mirinae
- Subfamily Bryocorinae

Here the subfamily Isometopinae contains all the monophyletic groups below it, and represents the branch of the tree that includes the branches above. The subfamily Psallopinae is part of Isometopinae, but contains all those groups below it, and represents the branch that includes all the branches above it (Wiley and Lieberman 2011, 237). On this sequencing convention, the taxa all have the same subfamily rank, but that rank does not indicate a particular classificatory level relative to the phylogenetic tree. The advantage to this convention is that it will not be necessary to add the many additional ranks to represent the tree of life. This solution is only partial, however, and can generally be used only on asymmetric trees, in which sister groups at each level have different numbers of included taxa. Here, for instance, the two sister groups in Isometopinae have one (Isometopinae) and five taxa (Psallopinae, Phyllinae, Cylapinae, Mirinae, Bryocorinae) respectively (Wiley and Lieberman 2011, 237).

Several solutions to the ranking problem have been proposed that aren't based on the Linnaean framework. One proposal is based on the idea that indentation can indicate relative inclusivity and the branching of the phylogenetic tree. On a pure version of this approach (which doesn't include any Linnaean ranks), the same hierarchical classification of Miridae would be represented as below:

Miridae
    Isometopinae
        Psallopinae
            Phyllinae
                Cylapinae
                    Mirinae
                        Bryocorinae

Each level of indentation indicates the level of inclusivity in the classification, and the branching order on the phylogenetic tree. Bryocorinae, for instance, is a subgroup of Mirinae, which is a subgroup of Cylapinae.

The indentation system has been combined with a naming system in an approach known as "Phylocode," advocated by Kevin de Queiroz, Jacques Gauthier, and others (de Queiroz and Gauthier 1992, 1994; de Queiroz and Cantino 2001). Here the names of monophyletic groups, or clades, are "defined" by indicating their particular places on the phylogenetic tree through reference to the members of a clade. A definition of a taxon name might, for instance, be something like: "the most recent common ancestor of archosaurs and lepidodaurs and all of its descendants" (de Queiroz and Gauthier 1992, 462). The traditional names of the taxa, such as 'archosaurs' and 'lepidosaurs,' could be retained, but the traditional Linnaean suffixes would no longer indicate rank. And the familiar genus-species binomial could also be retained, but without the first name indicating genus rank. There are many complications to this approach, and so far it has yet to achieve widespread acceptance.

A pure indentation system that doesn't use Linnaean ranks has the advantage that a revision of the phylogenetic tree doesn't *automatically* require a rank or name change. It merely requires a change in the relative indentation in the classification. So modifications can be made with relatively few costs. On the other hand, the indentation convention may work well when just a small portion of the tree of life is under consideration and consequently there are few indentation levels. But if a large branch on the tree of life, or even the tree of life itself, is being represented, it isn't clear how to represent and keep track of the many thousands (at minimum) of levels of indentation. It would take many pages to represent the tree of life. And if so, some method of comparing indentation levels across multiple pages would be required.

Another proposal, developed by Willi Hennig and others, is to indicate location on the phylogenetic tree by the use of numerical prefixes. Hennig illustrated this proposal by his classification of Mecopteroidea (scorpionflies), a taxon with two sister groups, Antilophora and Amphiesmenophora, each of which contains two terminal groups – Diptera and Mecoptera in the former and Lepidoptera and Trichoptera in the latter. Hennig assigned the numerical prefix 2.2.2.2.4.6 to Mecopteroidea to indicate its place in Insecta. The two sister groups within Mecopteroidea, along with their four terminal groups, were then assigned additional numbers indicating phylogenetic relations (Wiley 1981, 201):

2.2.2.2.4.6 Mecopteroidea
2.2.2.2.4.6.1 Amphiesmenophora
2.2.2.2.4.6.1.1 Trichoptera
2.2.2.2.4.6.1.2 Lepidoptera
2.2.2.2.4.6.2 Antilophora
2.2.2.2.4.6.2.1 Mecoptera
2.2.2.2.4.6.2 2 Diptera

This system, like the indentation system, has the advantage that the taxon names need not be changed if the phylogenetic tree is revised. And a change in the classification is relatively easy, requiring only a change in the numerical prefix. There are some obvious disadvantages, however. First the numerical prefix is likely to get very long as it represents bigger branches on phylogenetic trees. The preceding classification begins with the Insecta branch in the tree, and contains only the numerals indicating placement within that taxon. But what if it began with the eukaryote branch? The numerical prefixes for this group classified earlier would then likely be so long as to become difficult to even list, let alone compare with other taxa. Moreover, although these numerical prefixes might have some advantages for computers, they are difficult for humans to remember and compare (Wiley and Lieberman 2011, 245–247).

Fossils and Ancestors

A second complication in basing classification on the tree of life is the placement of fossils. Fossils are problematic, first, because there is generally much less information about their morphology, behavior, and molecular characters. But also one might think that the fossils we find in the geologic strata are at least sometimes members of species that were ancestors of living groups. If so, we might then think that these fossil species should be placed at a node or somewhere on a branch below the tip. Trees of the past have often done so. *Homo erectus*, for instance, was commonly placed at a node or on the branch leading up to *Homo sapiens*. But on the currently standard approach to tree building, there is no method for identifying fossils as branch nodes or on the branches below the tips. Fossils are generally just treated like extant organisms to be analyzed for phylogenetic relationship. Baum and Smith explain in their *Tree Thinking*:

> While some fossils might be actual ancestors of living species, the best way to approach fossils is to think of them as organisms that were collected in the present (as indeed they were), but stopped evolving a long time ago. That is to say fossils are best viewed as tips of the tree that have a shorter branch (in units of time) connecting them to the (inferred) ancestors. They are treated as living forms that have undergone no evolution in the millions of years since they were entombed in rock. (Baum and Smith 2012, 41)

Fossils are always placed at the tips because they are analyzed in terms of their shared derived traits (synapomorphies) and placed into sister groups on that basis, using the same parsimony or statistical methods used to place extant taxa on the tree. *If* the tree has a time scale, then the tips of the fossil branches may be placed lower in the tree than extant species, but still at a "crown." If the tree does not have a time scale, then the fossil species will be placed just as if they were still living organisms.

This convention for placing fossils at the crown of the tree of life is obviously misleading though, insofar as it treats extant and extinct species identically. Some systematists try to remedy this by placing the fossil taxa on the tree, and indicate their status by applying the rankless label 'plesion.' Other systematists indicate fossil taxa by denoting it with the dagger symbol (Wiley and Lieberman 2011, 237–238). But if a fossil species is an ancestor of *any* living species, and if the tree of life is intended to represent the actual evolutionary history, then surely it should be at a node or a branch lower on the tree. This can be indicated, even if it is not known where the fossil belongs on the tree and in a classification that represents that tree. One convention that does so, places the fossil taxon in the classification, but indicates that its place is uncertain by the label 'sedis mutabilis,' as Wiley does in the classification that follows. This indicates the fossil taxon *Neopterygius primus* has some indeterminate relationship to the two infraclasses Holostei and Teleostei.

Subclass Neopterygii (higher bony fishes)
 Neopterygius primus, sedis mutabilis
 Infraclass Holostei, *sedis mutabilis*
 Infraclass Teleostei, *sedis mutabilis*

If *Neopterygius primus* were taken to be the ancestor of these two infraclasses, then it could be treated as the stem species of the group, and identified with the entire group, as Hennig proposed.

> From the fact that … the boundaries of a "stem species" coincide with the boundaries of the taxon that includes all of its successor species, it follows that the "stem species" itself belongs in this taxon. But since, so to speak, it is identical with all the species that have arisen from it, the "stem species" occupies a special position in the taxon. If, for example, we knew with certainty the stem species of the birds (and it is only from such a premise that we can start I theoretical considerations), then we would no doubt

> have to include it in the group "Aves." But it could not be placed in any of the subgroups of Aves. Rather, we would have to express unmistakably the fact that in the phylogenetic system it is equivalent to the totality of all species in the group. (Hennig 1966, 71–72)

If so, then it could be classified in the hierarchy by placing it within parentheses next to the relevant taxon:

Subclass Neopterygii (*Neopterygius primus*)
    Infraclass Holostei, *sedis mutabilis*
    Infraclass Teleostei, *sedis mutabilis*

Wiley argues that this convention has the advantage that we can incorporate stem species into the classification without too much risk, even though the evidence may be sparse. The classification does not have to be substantially revised if *Neopterygius primus* isn't the ancestor (Wiley 2011, 243).

To some degree the problems with fossils and ancestors are unavoidable. We simply cannot know as much about extinct taxa as extant taxa, and that makes their placement in the tree of life and classification uncertain. But the methods used by phylogeneticists also create problems. By grouping just on the basis of synapomorphies into monophyletic clades, it isn't clear how to distinguish fossil taxa from extant taxa on the tree. To place them lower on the tree, at nodes or lower branches, some other method is required. It is not yet clear what the best approach to this would be. And it is not clear how to determine ancestral species in general. The standard method seems to assume that they simply have the ancestral, plesiomorphic traits of all its descendant groups. So we can think about a *hypothetical* ancestor as having just those ancestral traits. But that alone doesn't allow the inference that any actual taxa are the ancestors of a particular group. As Alessandro Minelli observes, "Although ancestors are or have been actual organisms, rather than abstractions, they cannot be recovered by cladistic methods" (Minelli 1993, 37).

Gene Trees

So far we have focused primarily on species trees that represent the branching evolution of species, as they split and form new species. We could also represent evolutionary history in terms of character evolution through character trees, which represent the origin and branching of new characters traits. As we saw in the first part of this chapter, many of the species trees of the past often had associated character trees that tracked character evolution – the formation of new traits, along with the origin of new species. William King Gregory, for instance, drew character trees that showed the evolution of primate hands and feet, the right humerus in mammals, cranial morphology in dogs and their relatives, and tooth evolution in bears (Pietsch 2012). The basic idea behind these character trees is that as new species form and change, new characters appear and change, and this branching character evolution largely corresponds to the branching evolution of species taxa. But notice we can also think about this branching evolution in terms of the genetic basis for the development and inheritance of these characters. We can represent evolutionary history not just by the characters produced in development, but by the genes that guide development. Gene trees can be constructed that depict the evolution of genes.

These three kinds of trees – species trees, character trees, and gene trees – all represent evolutionary history. But they don't necessarily represent it in the same way, with the same structure and branching order. With each new speciation event, branches appear in the species tree, but not all character traits or genes change with the formation of a new species. For those characters or genes that don't change, there will be no branching on the character or gene tree. So although some character or gene trees may exhibit the same branching structure as the species tree, others may not. These differences in the trees do not necessarily challenge the idea that we can base biological classification on the species tree. It merely implies that some character trees and some gene trees can be used to reconstruct the species tree, and some cannot. And if all organisms were asexual and uniparental, that would largely be the end of the story. But things get much more complicated in sexual, biparental organisms, through introgression and hybridization.

Introgression is when a limited number of organisms of one lineage or branch reproduce with members of another lineage. This results in a limited mixing of genetic material in the two populations. For example, it has been proposed that the presence of the *hypoxia pathway gene* in Tibetans, a gene that produces changes in hemoglobin concentrations and increased functioning in high altitudes, is a consequence of introgression by Denisovans into the *Homo sapiens* lineage (Huerta-Sánchez et al. 2014, 194–197). If this is right, then a few Denisovans interbred with members of the human lineage and exchanged genetic material. Here the interbreeding was insufficient to produce a complete mixing of genetic material, but the hypoxia pathway gene will have a gene tree that doesn't correspond with the strict species tree, in that *this* gene tree doesn't represent the human and Denisovan lineages as being separate. Because of this introgression, we might also think of the two branches on the species tree as joining or *reticulating*. The Denisovan branch has, to some small degree, joined with the modern *Homo sapiens* branch. If so, then the tree of life at this level is not exclusively branching, but is more like a web or network. But we could also preserve the exclusively branching tree structure by distinguishing the *dominant tree*, which ignores the introgression, from the *minor tree*, which represents the introgression. And if introgression went the other way as well, from the modern human populations into the Denisovans, there would be a second minor tree that represents the second direction of introgression (Baum and Smith 2012, 159). In this case, the two minor trees represent the gene histories from the genetic contribution of the interbreeding organisms in each population lineage. But the dominant tree also misrepresents the tree of life at this level in that it ignores the limited reticulation or joining of lineages through introgression. If we insist on representing this relationship with tree diagrams, we will need three trees rather than one.

In *hybridization* two population lineages interbreed on a much larger scale and fuse into a relatively homogeneous population, forming a new species. In these cases of lineage fusion, the reticulation of the branches is complete. Here the exclusively branching tree structure is even more misleading, and the weblike or network structure is more accurate in representing evolutionary history. We can maintain the branching structure, however, by recognizing two exclusively branching *codominant trees*, representing the roughly equal genetic contributions of the interbreeding organisms in each lineage. Strictly speaking, these codominant trees are inconsistent with each other though, representing different branching structures.

Introgression and lineage fusion are significant problems in that they are relatively common in evolution, particularly in populations that have recently diverged. There might be *intraspecific* hybrids, with interbreeding between two subspecies. Or there might be *interspecific* hybrids, between two closely related species, such as the "liger," a hybrid between lions and tigers. At these levels the reticulation doesn't cross multiple branches of the tree of life, but occurs mostly within a single branch and at the branching node. Higher-level hybrids, *intergeneric* - between different genera, or *interfamilial* - between different families, are much less common among animals. And generally the interbreeding at anything but the lowest levels results in sterile offspring. So there would be no new enduring branches on the tree. But there is also significant lineage fusion among plant species that are not closely related. According to analysis by James Mallet, at least 25% of plant species have some involvement with hybridization and introgression (Mallet 2005). Consequently the fusion or reticulation of branches in the plant portion of the tree of life is much more common. And the representation of that part of the tree of life as exclusively branching is much more misleading. But Mallet also suggests that 10% of animal species involve hybridization or introgression. If so, then in animals as well as plants, the web or network metaphor may in some places be more accurate than the tree metaphor in representing evolutionary history.

But even on the branches of the tree where there is no introgression or lineage fusion, the tree of life can misrepresent evolutionary history for sexual biparental organisms. Gene trees can conflict with species trees because of *incomplete sorting* and *deep coalescence*. In diploid sexual organisms, there are two copies of a gene at each position or locus. Each organism inherits one copy from its mother and one from its father. These copies may be the same, if they have the same nucleotide composition, or they may differ and be two different alleles or versions of that gene. Each organism can therefore potentially have two different alleles at each locus, depending on the genes inherited from each parent. Since there are two copies, one inherited from the mother and one from the father, there are then potentially four different combinations of versions of each gene, or alleles, at each site.

Because in sexual biparental organisms, each offspring receives one copy of each gene but not the other, from each parent, two of the parental copies are not passed on in reproduction. And some of these copies are inherited in more than one offspring. The histories of the alleles, or different version of the same gene, are therefore treelike, with branches where a single allele is inherited by two or more offspring, and tips where the allele is not passed on to any offspring. In a group of organisms this leads to sorting, where at some point in time, some of the allele lineages in a population go "extinct," by virtue of not being represented in any of the offspring. *Incomplete sorting* is where there is more than one allele lineage in a population over time. If these different allele lineages persist over speciation events, then we can get *deep coalescence.*

We can understand *coalescence* best by starting with a current generation of individuals and the alleles of that generation. Baum and Smith explain:

> If you pick any two alleles in the current ... generation, you could trace backward down their lineages until they converged on a common ancestral allele. At the point when the two gene lineages reduce to single common ancestor, they are said to undergo coalescence. Some alleles from the current generation coalesce only one generation in the past, while other pairs coalesce only at the last common ancestor shared by all alleles in the current generation. (Baum and Smith 2013, 142)

So coalescence in two different alleles, then, is the point at which there is common ancestry for these two alleles. Deep coalescence is when the common ancestor of two alleles, whose lineages both persisted in incomplete sorting, precedes the speciation event that produced two or more new species. This is illustrated in Figure 5.2.

Here there are three gene lineages, 'a,' 'b,' and 'c,' represented by the thin black and gray lines, within three species lineages (A, B, and C). In this tree, species C branched first, whereas species A and B branched afterward. But the gene lineage 'a' branched *before* 'b' and 'c.' This deep coalescence is the result of incomplete sorting of 'a' and 'b.' Both persisted in a single lineage of organisms until that lineage itself later split into two species, with the 'a' gene lineage found in just one species while the 'b' lineage was found in just the other. The important point here is that the species tree would have a branch with just A and B on it. But the gene tree would instead have a branch with just b and c. This species tree has a different branching order than the gene tree. This suggests that in some cases species trees may not be accurately representing something important about evolutionary history, the evolution of genes.

Figure 5.2. Deep Coalescence.
(Courtesy of Alana Baldwin, Multimedia Services, University of Alabama.)

Horizontal Gene Transfer

Another challenge to the standard tree thinking in classification comes from lateral or *horizontal gene transfer* (HGT). In introgression and hybridization, genetic inheritance is vertical, in the sense that it is passed from parent to offspring. So even if two branches of the tree of life merge in hybridization, the transfer of genes is still vertically from parent to offspring along the branches rather than horizontally across branches. In sexual species there is merely the merging of two vertically transferred genomes. But genetic material can also be transferred horizontally - across the branches. This is most common in prokaryotes, single-celled organisms that lack a membrane-bound nucleus, and that include Bacteria and Archaea.

There are three widely recognized processes of horizontal gene transfer in prokaryotes. In *conjugation*, DNA is transferred by plasmids, short segments of extrachromosomal DNA in cells, across a cytoplasmic bridge from one cell to another. In *transduction*, DNA is transferred by *phages*, viruses that infect prokaryotes. And in *transformation*, cells take up free-floating, "naked" DNA in the immediate environment. The success of these mechanisms of HGT is dependent on physical proximity in a common environment, effectiveness of barriers to integrations, and how well the DNA can be functionally integrated in the new genome (Bapteste and Burian 2010, 714). But there is also evidence that HGT often succeeds in these prokaryotes.

In one estimate, 81% of the genes examined in 181 complete prokaryotic genomes were laterally transferred (Dagan et al. 2008). If this is accurate, HGT is hardly trivial in the tree of life. After all, the first half or two-thirds of life's history was prokaryotic, and much of life today is prokaryotic.

But even among eukaryotes – organisms with a membrane-bound nucleus, HGT seems to be common. This would be most expected among *protists*, single-celled eukaryotes that include green and red algae, protozoa, and slime molds. And it has been increasingly found in this group. In fungi, for instance, there seems to be a significant amount of HGT with other fungal lineages, prokaryotes, and other multicellular eukaryotes (O'Malley 2010, 537–539). Plants also seem to have acquired genetic material from prokaryotes by HGT, as have animals. Maureen O'Malley lists some instances:

> Some plant-parasitic nematodes have acquired bacterial genes that enable the nematodes to modify plant cell walls, thereby damaging the plant, but nourishing the nematode ... The sea squirt, *Ciona intestinalis*, one of the few animals able to synthesize cellulose, gained this capability from the lateral acquisition of cellulose synthase genes from bacteria ... Even more peculiarly, eukaryote-to-prokaryote transfers of tubulin and actin genes have given the acquiring bacteria totally novel structural resources. (O'Malley 2010, 543)

While O'Malley cautions that some of the stories of HGT are likely exaggerated, she also suggests that these discoveries are "only scratching the surface" (O'Malley 2010, 544). Researchers have recently looked at twenty-six animal species, including ten primate, twelve fly, and four nematode species. In these species they discovered that bacteria and protists were the most common donors of genetic material, although there were apparently some viral donors as well. In humans they found genes acquired by HGT related to amino acid and lipid metabolism and immune response, and more. These researchers conclude that in eukaryotes HGT seems to be common:

> Although observed rates of acquisition of horizontally transferred genes in eukaryotes are generally lower than in prokaryotes, it appears that, far from being a rare occurrence, HGT has contributed to the evolution of many, perhaps all, animals and that the process is ongoing in most lineages. Between tens and hundreds of foreign genes are expressed in all the animals we surveyed, including humans. (Crisp et al. 2015)

As this suggests, there is at least some evidence that HGT is common in humans. On one estimate 8% of human DNA was added to the genome by HGT solely from retroviruses. (Bapteste and Burian 2010, 716). If these estimates of HGT are anywhere near the true magnitude, even in the absence of introgression or hybridization, the branches of the tree of life have many horizontal connections. If so, any tree of life that represents only the vertical genetic transmission from ancestor to descendant along the branches is clearly missing an important part of evolutionary history.

Rethinking the Tree of Life

As we have seen, there seem to be major problems with the tree of life. First is the challenge it poses to Linnaean ranking. If the branches of the tree are taken to imply ranking, then the many thousands, perhaps millions of branches cannot possibly be reflected in the mere twenty or so Linnaean ranks. This seems to be more of a problem though for the Linnaean ranking system than for the tree itself. If the basis for classification should be evolutionary history, as Darwin argued, and that history is best represented as a tree, then what is needed is just a better way to represent the tree in classification. The tree isn't the problem.

But as we have just seen, there also seem to be problems with the tree of life itself. The first *substantive* problem with the tree is that with current methods it isn't clear how to incorporate fossils and identify ancestors. If fossils are treated the same as extant taxa they will be at the tips of the branches, even if to best represent evolutionary history they should sometimes be placed lower on the branches or at a node. Second, genes seem to form trees just as species do. In some sense the vertical transmission of genes underlies the species trees. After all, what is important in ancestor–descendant relations is the vertical transfer of genes. But because of incomplete sorting and deep coalescence, gene trees will often be inconsistent with each other and with species trees. One gene tree will group species one way, and another gene tree will group them differently, and both might be inconsistent with the species tree. This implies that no single tree will adequately represent evolutionary history. Third, introgression and hybridization suggest that the exclusively branching structure of the tree of life misrepresents evolutionary history. Some of the branches reticulate – rejoin other branches. Finally, an exclusively branching tree does not seem able to represent HGT, which would require many, almost innumerable, cross connections across branches. And at the exclusively prokaryotic base or trunk of the tree of life, the connections are likely web-like with little branching structure. The bottom line is that a single tree of life cannot adequately represent the many important processes involved in evolutionary history, and by representing that history as a strictly branching structure, it is not just inadequate, but also misleading.

How should we respond to these problems with the tree of life? The ranking problem is perhaps the least serious in that it can be solved not by abandoning or modifying the tree of life, but by abandoning or modifying the Linnaean conventions. There are substantial costs to either of these solutions though, as a new system must be implemented. That might require learning new names and forgetting the old, as well as working out the details of a modification to, or replacement for the Linnaean ranks. But the other problems are more substantive. They are problems with the tree of life itself. One solution is to reject the tree of life altogether, because it doesn't accurately represent evolutionary history. It neglects fossils and ancestors, discordant gene trees, introgression and hybridization, and HGT. (See Doolittle 2010 and Franklin-Hall 2010)

Another possible response is to turn to *forest thinking*. Perhaps many of the problems with the tree of life can be solved by the recognition of multiple trees. On this response there is no single tree of life, rather there are many trees that represent different processes. If we allow for multiple discordant gene trees, such as the minor trees that show introgression and the codominant trees that show hybridization, perhaps we are getting closer to representing the true history of evolution. But it still isn't clear how the branching tree structure can represent horizontal gene transfer. For this we need a web or network. It doesn't look like forest thinking by itself will suffice, but perhaps networks are in the forest, as well as trees. Moreover, what attitudes should we have toward the different trees in the forest? Do they all coexist as they would in a real forest? Forest thinking, by itself, doesn't give us obvious and clear answers to these questions.

A third response is *modified tree thinking*. Perhaps we can modify a single tree of life to show introgression and hybridization, as well as HGT. This tree would not be exclusively branching. In fact the first half of life's history, which was exclusively prokaryotic, would have much more of a web structure, as genes were exchanged horizontally from one cell to another.

We could think of this as a large trunk with many horizontal connections along with the vertical. Above this trunk there would be some branches that rejoin with other branches, either wholly or in part. But even above the trunk there would be extensive reticulation. Perhaps conventions could be used to distinguish introgression from hybridization and HGT. And perhaps to represent the vertical histories of genes, we could treat the branches not as single undifferentiated structures, but more like ropes, with various genetic strands traveling along the branch and sometimes diverging from the main branch. These genetic strands could sometimes branch at different places on the tree than where the ropes themselves branch. This would be an immensely complex tree, but it could in principle reflect many of the complexities of evolutionary history. And it may not be as complex and chaotic as feared. Since horizontal gene transfer is biased – it occurs more in closely related groups – there may be more coherence in the tree of life than feared (Andam et al. 2010, 599). Many of those who think trees to be important have responded in this way, by modifying trees, at least in some limited fashion, representing some of these processes in limited sections of the tree of life. Baum and Smith do this in their *Tree Thinking* (2013), as we see in the tree in Figure 5.2 that shows the transmission of alleles within a lineage

A fourth response is methodological. It is to treat trees as *models* that represent one or more features of evolutionary history but not others. On this approach, we construct trees with specific purposes in mind, and to serve those purposes. Joel Velasco argues for this approach, calling it the "modeling defense of trees."

> We can call the modeling defense of trees the view that trees are useful because they are very good models. Models contain idealizations. Roughly, idealizations deliberately simplify or alter something complicated in order to better understand it or to better understand something else. (Velasco 2012, 629)

On this approach, the mere fact that the tree of life misrepresents evolutionary history is not a problem. *All* models misrepresent the world. Galileo's model of the pendulum, Kepler's model of the solar system, Bohr's planetary model of the electron, and the double helix model of DNA all have something in common. They all misrepresent the things they model. But these misrepresentations are nonetheless useful for various reasons.

One way to misrepresent is simply to leave out the factors or elements that are too complicated and assumed to be largely inconsequential. Velasco calls this type of model a *minimal idealization*.

> Here, the model contains only the core causal factors that are relevant in the situation at hand. Obviously, there are many aspects of genealogy that a tree does not represent. If we have a tree representing the great ape species, we know that traits are actually inherited by individual organisms, not by species. But for at least some of uses of trees, these details are simply ignored as irrelevant to the problem at hand. (Velasco 2012, 629)

Sometimes these ignored factors do make a difference, but a difference that can be bracketed or set aside. Velasco argues that this is the case with the limited horizontal gene transfer in diverging lineages. Among *some* organisms, that gene transfer is limited and can be ignored. Notice, however, that this may not be true of all organisms. In these more problematic cases the tree model may not be useful.

Another type of model, according to Velasco, is the *Galilean idealization* that introduces deliberate distortions. He illustrates this with Galileo's assumptions that the Earth is flat and that a rolling ball is perfectly spherical. Here, the model deliberately misrepresents some factors or elements, rather than excluding them. It isn't clear though that this is a distinctive kind of model. Galileo could be understood as just ignoring the curvature of the Earth and the spherical deviation, treating them as unimportant, rather than introducing distortions.

A third type of model identified by Velasco is the *multiple models idealization*. The idea behind this approach is that we can construct multiple related but incompatible models, each representing different things and relying on different assumptions. He draws an analogy with maps: "Some maps of New York City focus on the layout of the streets and ignore topographical features, others carefully note changes in altitude, but would not help if you were lost trying to find the library" (Velasco 2012, 631). The important idea is that maps suit pragmatic interests. For hikers, a good map would typically represent trails, elevations, rivers, mountains, meadows, and so on. But for drivers, a good map would typically represent roads, bridges, tunnels, and parking lots. The fact that a hiking map doesn't represent parking lots and roads does not normally count against that map – for hikers anyway. Similarly, a road map is not normally criticized for neglecting meadows and trails. In fact we might be inclined to criticize a road map for including things not relevant to driving, and a hiking map for including things not relevant to hiking. Moreover, and in contrast to both the hiking and driving maps, we might instead be interested in geologic formations, underground aquifers, earthquake faults, and more. If so, then the optimal maps will represent just these things.

We can think about trees on this map analogy. What are we interested in? Species? Individual organisms? Cells? Genes? Horizontal gene transfer? Each one of these interests has implications for what sort of tree diagram or network will be of most use. If, for instance, we are most interested in sexually reproducing, large vertebrate *species*, then an exclusively branching phylogenetic *species* tree might be just what we are looking for. On the other hand, if the bacterial genes that affect virulence and that sometimes transfer horizontally are what interest us, then a strictly bifurcating species tree would likely not be of great utility. A web would be of more value in this case. Or perhaps we might be interested in the history of particular disease genes. Here a gene tree would be most useful. And if we wanted to know what species these genes were found in, we might need a tree that shows both species lineages and gene lineages. On this way of thinking about trees as models, what is properly represented depends on interest and facts about the world.

But this way of thinking about phylogenetic trees may seem unsatisfactory to those building a single grand tree of life. The goal here is not typically taken to be one big misleading model of species histories. Rather the intention seems to be the representation of some fundamental pattern of evolution. If useful models were the only desired outcome, why try to construct a single tree of life? This grand tree of life would be a huge, unnecessary, and misleading project. Is there some use for this tree of life beyond the value of trees as models? Lurking here is a psychological function of a tree of life. In his *Origin*, Darwin used his tree diagram heuristically, to help his readers better understand his ideas. In chapter IV, for instance, he used it to illustrate his principle of divergence. According to this principle:

> The modified descendants of any one species will succeed by so much the better as they become diversified in structure, and are thus enabled to encroach on places occupied by other beings. Now let us see how this principle of great benefit being derived from divergence of character,

> combined with the principles of natural selection and extinction will act. (Darwin 1859, 116)

In the next sentence Darwin refers to his tree diagram: "The accompanying diagram will aid us in understanding this rather perplexing subject" (Darwin 1959, 116). The details of how this principle of divergence was supposed to work are not important here, but what is important is the psychological function of the tree in getting the readers of the *Origin* to understand a key evolutionary process.

The tree diagram was similarly used in chapter XIII of *The Origin*, to help the readers understand how evolutionary processes could produce a group-in-group classificatory structure:

> I request the reader to turn to the diagram illustrating the action, as formerly explained, of these several principles; and he will see that the inevitable result is that the modified descendants proceeding from one progenitor become broken up into groups subordinate to groups. (Darwin 1859, 412)

This heuristic use of the tree could then help the readers understand how naturalists had come to classify species into genera, genera into families, families into orders, and orders into classes:

> So that we here have many species descended from a single progenitor grouped into genera; and the genera are included in, or subordinate to, sub-families, families, and orders, all united into one class. Thus, the grand fact in natural history of the subordination of group under group, which, from its familiarity, does not always sufficiently strike us, is in my judgment fully explained. (Darwin 1859, 413)

Darwin seemed to think that a *full* explanation required more than just an appeal to a cause, law, or principle, but also depended on something more – something subjective and psychological. One way to do this is to *show* how the important processes might work, so the reader could *see* it. The tree diagram was in this sense a heuristic aid so that the readers of *The Origin* could *see* how biological classification, as they knew it, could be a consequence of the divergent branching processes of evolution. Contemporary advocates of tree thinking often find this same subjective psychological value. Baum and Smith, for instance, focus on this in the preface to their book *Tree Thinking*:

> Tree thinking is not just a practical skill to be learned, like riding a bicycle or doing long multiplication. It certainly is an important tool for the biologist, but it offers much more than that. The evolutionary perspective offered by tree thinking helps one appreciate biological diversity in much the same way that some knowledge of music theory can help one appreciate a great symphony. (Baum and Smith 2012, xv)

The many problems with the evolutionary tree don't obviously undermine this psychological value.

This may be a more general feature of science. Explanations that are associated with an iconic image seem to resonate with us in a way that other explanations may not. The history of science is permeated by simplified images of atoms, molecules, solar systems, and the universe itself - each of which misrepresents the complexity of nature and has its own limitations. One striking and iconic image from twentieth century biology is of the double helix of the DNA molecule. This image has come to symbolize a reductionist explanation of development - how a single molecule can be used to explain heredity and development. But the double helix image is also deeply misleading in that it neglects the causal roles of the environment and epigenetic processes in development. Nonetheless, the image of the double helix, like the evolutionary tree, is subjectively satisfying and engaging. For Darwin's readers and the evolutionists who have followed, the tree of life asks us to think about evolution in a particularly satisfying way, as underlying a grand narrative represented by the growth of the tree of life.

Perhaps the most reasonable stance here is pluralistic. This stance recognizes that we can use trees as models to do many things, perhaps to facilitate the understanding of evolutionary processes, or to represent particular processes, whether those processes be the divergence and diversification of species, or of characters or genes. Which processes are to be illustrated or represented depend on particular interests and the utility in serving those interests. So ultimately the value of the tree is at least partly pragmatic. A tree of life cannot represent everything accurately, any more than a single map could represent everything accurately, but like a useful map it might represent some things of value and interest. It might represent the patterns in evolutionary history that interest us the most or at a particular time. How to best do that might is an open question. But whatever the case, it is clear that no single tree of life will serve all of our purposes or answer all of our questions.

6 The Species Problem

The Species Category

In Chapter 5 we looked at tree thinking in classification, its virtues, and some of its problems. Among those problems is the *ranking problem*. If classification is to be purely phylogenetic and based on the branching of the tree of life, and if the branches are to be represented in classification by ranks, then the Linnaean system with its mere twenty to thirty ranks is clearly inadequate. Many more ranks will be required. But even if we set this problem aside, there is another ranking problem in biological classification. One rank – the species rank – seems to have a unique status. Species seem to be different in important ways from the higher ranks – genera, classes, orders, and so on, but what constitutes species and makes them different has been debated for a long time. In this chapter we will look at the philosophical dispute over the nature of species.

There are several reasons for the unique status of species. The first and most obvious is that the species level is the lowest universally recognized level in the hierarchy. In classification, organisms are grouped together into species in *microtaxonomy*, and then those species are grouped together into higher levels – genera, families, classes, orders, and more – in *macrotaxonomy*. The species level is then the first and lowest level of grouping. It is also the category usually represented by the tips of the branches in the tree of life.

While species are the lowest *universally* recognized rank, subspecies (or varieties) are also sometimes recognized. The rules of nomenclature allow for the recognition of subspecies through the use of a trinomial, generated by adding a third, subspecies name to the species binomials. For instance, multiple tiger subspecies have been named and recognized by some systematists, based on relatively minor but still apparent geographic variations: *Panthera tigris altaica* for the Siberian tiger, *Panthera tigris tigris* for the Bengal tiger, and *Panthera tigris sumatrae* for the Sumatran tiger. But even for those biologists and systematists that recognize subspecies by giving them names, the species rank has usually been regarded as having special significance. On one standard way of thinking, species groupings have a unique evolutionary significance, because species are reproductively isolated. Organisms of different species are much less likely to reproduce than organisms of the same species but different subspecies. This has obvious implications for divergent evolutionary change. Once reproductive isolation has been established, then two groups of organisms can evolve in divergent ways, because their differences would no longer be blended in reproduction.

The species level is often taken to have greater evolutionary significance relative to higher taxonomic levels as well. First, interbreeding within species (at least in sexually reproducing species) provides cohesion and gene flow among the members that is not found to the same degree at higher levels, in genera, families, and classes for instance, where there is much less interbreeding. Second, natural selection is usually assumed to operate similarly within species in a way that it doesn't at higher levels. Since members of a single species are more uniform in morphology and behavior, they will be subject to similar selection pressures that will in turn tend to maintain uniformity within the species. Members of higher level taxa - genera, classes, and orders - typically differ in terms of morphology and behavior, occupy different niches, have greater differences in selection pressures, and are more likely to diverge on these grounds. Finally, some theorists think there are other ways that species might function in evolution, perhaps through species selection, in ways that genera and higher levels don't. (See Okasha 2006 and Jablonsky 2008 for instance.)

Although the claims about species selection are controversial, and there is some reproduction across species boundaries in hybridization and introgression, there is still general agreement that the species category is special relative to evolutionary processes. Evolutionary change occurs through the formation of new species, the extinction of old, and the modification of species taxa by natural selection over time. For these reasons, species are commonly described as "the units of evolution" (Mayr 1982, 252; Dupré 1993, 42; Hull 1998, 295). In the introduction to a volume titled *The Units of Evolution: Essays on the Nature of Species*, Marc Ereshefsky expressed this view:

> Biological systematists attempt to provide a taxonomy of the world's organic diversity. Evolutionary biologists attempt to explain why that diversity exists. Species are often viewed as *the* evolutionary units of the organic world. The members of a species are exposed to common evolutionary processes that cause those organisms to evolve in step with one another or maintain a common evolutionary stasis. When such processes are disrupted, a new species may form. Higher taxa lack such processes. Thus any stability that a higher taxon has is the result of processes occurring at the species level. Similarly, any change that occurs in a higher taxon is due to a disruption of species level processes. Species are active agents in the evolutionary process, while higher taxa are passive aggregates consisting of species. (Ereshefsky 1992, xiii)

Because species are seen as evolutionary units, playing important roles in evolutionary processes, they are typically claimed to be real in a way that taxa in other categories are not (Panchen 1992, 333). This significance of the species category is implicitly endorsed by those who conceive the tree of life as a *species* tree.

This special status of species in classification and evolution is also reflected in how conservation biologists think about species. While it is common to say that biodiversity consists in all differences at all levels, from genes to species, genera, classes, orders, etc., and in differences of all kinds in both ecological and behavioral categories, species taxa seem to have a special status, and are often seen as fundamental units of biodiversity. In part this is because of the assumed role of species as evolutionary units. But also species are often thought of as gene pools, with the genetic resources for producing variation that can maximize and maintain fitness and ultimately viability. Viability, the ability to persist and adapt, is of crucial significance in conservation biology, and the species level, as a genetic pool of adaptive resources, seems to have special significance here.

The significance of species here is also partly due to the fact that speciation is often taken to be the origin of biodiversity. According to E. O. Wilson, for instance, a single species may change over time, but as long as it remains a single species, there has been no real increase in biodiversity. But in branching speciation, with its production of additional species, there is an increase in biodiversity (Wilson 1999, 51–52). If so, it is not surprising that estimates of biodiversity are often in terms of species counts. Richer, more biodiverse environments and ecosystems simply have more species (MacLauren and Sterelny 2008, 28; Maier 2013, 82). Based on this idea, Wilson has quantified the amount of biodiversity on Earth in terms of a total number of recognized species, which in 1999 he estimated to be in the range of 1.5 to 1.7 million. He estimated the total number of species, both recognized and unrecognized, to be around 10 million (Wilson 1999, xiv). If species counts are a good measure of biodiversity, it is not surprising that loss of biodiversity is also typically measured in terms of species extinctions. Moreover, legislative efforts to preserve biodiversity have also focused on species, notably in the endangered species legislation of the United States, including the Endangered Species Preservation Act of 1966, the Endangered Species Conservation Act of 1969, and the Endangered Species Act of 1973.

There are many complications to these claims that species are the units of classification, evolution, or biodiversity, nonetheless there is a longstanding and common tendency to think that species are real in *some* way or other that the lower and higher ranks are not. If so, what makes the species rank different? The standard answer in terms of interbreeding isn't fully satisfactory because many species reproduce asexually. An adequate answer to this question might involve saying *precisely* what species are, and what higher and lower level taxa are. Then we could see what is distinctive about species. Unfortunately, there is no consensus on the nature of species. As we shall see, there are multiple, inconsistent ways of thinking about species. In other words, there are different species concepts in use that divide up biodiversity in different ways. But as we shall also see, no single way of thinking about species seems adequate. The multiple ways of defining species presents a challenge to any biologist who, first, thinks that species are real, and second, relies on species groupings in research, as Joel Cracraft explains:

> The primary reason for being concerned about species definitions is that they frequently lead us to divide nature in very different ways. If we accept the assumption of most systematists and evolutionists that species are real things in nature, and if the sets of species specified by different concepts do not overlap, then it is reasonable to conclude that real entities of the world are being confused. It becomes a fundamental scientific issue when one cannot even count the basic units of biological diversity. Individuating nature "correctly" is central to comparative biology and to teasing apart pattern and process, cause and effect. Thus, time-honored questions in

> evolutionary biology – from describing patterns of geographic variation and modes of speciation, to mapping character states or ecological change through time, to biogeographic analysis and the genetics of speciation, or to virtually any comparison one might make – will depend for their answer on how a biologist looks at species. (Cracraft 2000, 6)

In this chapter we will look at this *species problem* – the use of multiple inconsistent species concepts, its implications, and possible solutions.

The Species Problem

In the title of his *On the Origin of Species*, Darwin seemed to endorse the special status of the species category. After all, he didn't title it "On the Origin of Genera," or "On the Origin of All Biological Taxa." But surprisingly he also seemed to doubt the reality of species. At the beginning of his chapter on variation in nature he wrote: "From these remarks it will be seen that I look at the term species, as one arbitrarily given for the sake of convenience to a set of individuals closely resembling each other" (Darwin 1859, 52). And at the end of the *Origin* he made a similar claim: "we shall have to treat species in the same manner as those naturalists treat genera, who admit that genera are merely artificial combination made for convenience" (Darwin 1859, 485). Part of the problem, in Darwin's view, was that naturalists adopted different definitions of the species category. They were conceiving *species* in different ways. He made this clear in the first sentences of his chapter on Variation under Nature in his *Natural Selection*, a book he was working on before writing the *Origin*:

> In this Chapter we have to discuss the variability of species in a state of nature. The first & obvious thing to do would be to give a clear & simple definition of what is meant by a species; but this has been found hopelessly difficult by naturalists, if we may judge by scarcely two having given the same. (Stauffer 1975, 95)

Here he lamented the fact that naturalists in his time were giving multiple, seemingly inconsistent definitions of species. Some were treating species taxa as groups of similar organisms, others as members of a parent-offspring lineage, others as reproductively isolated groups; others were associating species with a particular abstract type, and yet others were treating species conventionally, as just what naturalists identify as species.

The species problem in Darwin's time was, as it is now, that there were different, inconsistent ways of conceiving species, and no single species definition seemed adequate.

There has been an extensive debate about Darwin's views and whether he regarded species as real or not. (See, for instance, Stamos 2003, Wilkins 2009, Richards 2010, Mallet 2013.) But whatever his views were on this question, a century later the species problem remained. In 1942, Ernst Mayr outlined five basic concepts in use: practical, morphological, genetic, sterility, and biological (Mayr 1942, 115–119). He advocated the *biological*:

> A species consists of a group of populations which replace each other geographically or ecologically and of which the neighboring ones intergrade or interbreed wherever they are in contact or which are potentially capable of doing so (with one or more of the populations) in those cases where contact is prevented by geographic or ecological barriers ... Or shorter: species are groups of actually or potentially interbreeding natural populations, which are reproductively isolated from other such groups. (Mayr 1942, 120)

Forty years later Mayr gave three formulations of this *biological species concept*. The first was based on actual or potential interbreeding.

> Species are groups of actually or potentially interbreeding natural populations which are reproductively isolated from other such groups. (Mayr 1982, 273)

The second was in terms of reproductive communities:

> A species is a reproductive community of populations (reproductively isolated from others) that occupies a specific niche. (Mayr 1982, 273)

The third version was in terms of the reproductive isolating mechanisms, defined as follows:

> Isolating mechanisms are biological properties of individuals which prevent the interbreeding of populations that are actually or potentially sympatric. (Mayr 1982, 274)

Whether or not these three formulations are equivalent, Mayr thought this "biological" approach was better than the other ways of thinking about species for a theoretical reason. Reproductive isolation is necessary for divergent evolutionary change. But he recognized nonetheless that other ways of thinking about species were necessary because the biological species concept could not always be applied, in particular to geographically isolated groups of organisms (Mayr 1942, 120). These groups are a problem for the biological species concept because unless the individual organisms are in contact, we cannot tell whether they are able and inclined to reproduce.

Mayr's biological species concept has perhaps been the approach most widely used and taught in introductory biology courses. But it has an obvious flaw. It is based on sexual reproduction, but much of the biodiversity we find in nature does not reproduce sexually. So the biological species concept can apply to only a relatively small portion of life. There are several possible responses here. We might respond as Theodosius Dobzhansky did, and just deny that nonsexually reproducing organisms form species (Dobzhansky 1937, 319). If so, only a portion of biodiversity comes grouped into species. But few have followed Dobzhansky's lead, and for good reasons. There are obvious questions raised by his response, questions not easily answered. How should we then classify asexually reproducing organisms? Do we need a different basal category in our classifications? Perhaps "pseudo-species"? Is there another category we could place them in that would have equivalent evolutionary significance? Or perhaps there is simply no comparable evolutionary unit among asexual species. Moreover, if we cannot preserve asexual species in our conservation efforts – because there are none, what should we be preserving? Anyone who rejects the application of the species concept to asexually reproducing organisms would need to give satisfactory answers to these questions. Unsurprisingly, most theorists have instead looked for different ways to conceive species that could apply to asexual organisms as well as the sexual.

In the time since Mayr first developed his ideas about the *biological species concept*, species concepts have proliferated. Richard Mayden (1997) and Jody Hey (2001) have each compiled a list of more than twenty species concepts in use. While neither Mayden nor Hey classify species concepts this way, there seem to be three main types of concepts. First are the *process* concepts, grounded on various biological and evolutionary processes. The basic idea here is that we can identify and individuate species taxa on the basis of the biological and evolutionary processes that are responsible for the development or maintenance of species. Here we find Mayr's biological species concept based on sexual reproduction. Roughly, species are groups of organisms connected by reproduction and gene flow, but reproductively isolated from other organisms. But for those organisms that don't reproduce sexually, we have the *agamospecies* concept proposed to serve as an umbrella concept for all taxa that are uniparental or asexual (Mayden 1997, 389). But even for the organisms that sexually reproduce, there are other ways to think about species. For instance, we can conceive species not in terms of actual reproductive isolation, but in terms of the mechanisms that cause isolation. Hugh Paterson's *recognition species concept* does just that, and is based on the fact that organisms have specific mate recognition systems that consist in a set of signaling methods and physiological compatibilities that make sexual reproduction both possible and likely (Paterson 1993). This captures the idea that new species can form on the basis of the modification of mate recognition systems and that reproductive cohesion and isolation are in part products of these systems.

When we think about reproductive cohesion and isolation, we might also think about the underlying genetic basis. The *genetic species concept* follows up on this suggestion and is based on the idea that reproductive cohesion in a population is due to a common genetic foundation - the common gene pool that makes reproduction possible. Mayden attributed this concept to Dobzhansky, who identified species with "the largest and most inclusive reproductive community of sexual and cross-fertilizing individuals which share in a common gene pool (Mayden 1997, 399).

There are other processes that are responsible for the creation and maintenance of species, based on geographic isolation and natural selection. A *geographic species concept* is implicit in the discussions of geographically triggered speciation we find in Darwin, Dobzhansky, and Mayr (Richards 2010, 94, 107). Here species are geographically isolated and localized populations with distinctive divergent variations. The operation of natural selection is also a relevant process and plays a role in the *ecological species concept* proposed by Van Valen, which asserts that a "species is a lineage (or closely related lineages) which occupies an adaptive zone minimally different from that of any other lineages in its range" (Mayden 1997, 394–395; Van Valen 1992, 70). Adaptive zones are parts of the environment that exist independently of the particular organisms, but that maintain a common adaptive response within the group of organisms based on natural selection.

One problem with these process-based species concepts is that they are often difficult to apply. Just as we cannot easily tell whether geographically isolated groups of organisms could and would interbreed, it is often difficult to tell whether there are mate recognition systems that would prevent interbreeding. What can be applied in varying degrees, though, is the second kind of concepts, based on similarity. Most obvious is the *morphological species concept*, which asserts that "species are the smallest groups that are consistently and persistently distinct, and distinguishable by ordinary means" (Mayden 1997, 402). According to this approach, which is based on morphological distinctness, we can group organisms into species taxa based on the degree of difference in physical traits such as skeletal structures, sexual organs, size, shape, beak length, and more. This is a common criterion, in part because of its ease in application. But it also has its shortcomings. Most obviously, morphological distinctness is not able to identify and individuate the so-called *cryptic species*, groups of organisms that have no obvious morphological discontinuity or differences with other groups, but nonetheless cannot or will not reproduce with the members of these other groups. Moreover, there is often substantial variability within a species. The parasitic males of the barnacle species *Trypetesa lampas*, for instance, are much smaller than the females, lack gastrointestinal tracts, and live only long enough to fertilize the females. On a morphological criterion these males and females would surely be considered distinct species, and likely even different genera and families. But it is not just sexual dimorphism that is problematic. Developmental stages (such as we see in many insects) and environmental phenotypic variability often produce dramatically different morphologies. The larval stages of insects are very different from the adult, and plants of the same species typically vary dramatically between higher and lower elevations. All this morphological variability within what we take to be single species taxa, suggests that a purely morphological species concept will not do.

Another similarity-based concept is the *phenetic species concept*, which uses algorithms to establish overall similarity based on a set of variables (Mayden 1997, 404). This approach, as we saw in the discussion of phenetics in Chapter 4, has additional problem that different algorithms produce different groupings and there is no obvious reason to prefer one algorithm over another. The *polythetic species concept* is a "cluster concept that defines species in terms of significant statistical covariance of characters" (Mayden 1997, 408). Other similarity concepts are based on genes and DNA, such as the *genotypic cluster concept* and the *genealogical concordance concept*, which determines species groupings on the concordance of multiple independent genetic traits (Mayden 1997, 397).

All the similarity-based concepts seem to share the same basic problems. First, as we saw in Chapter 4, different systematists might group on the basis of different characters, and employ different similarity algorithms and clustering approaches. But it is not clear why we should prefer one set of characters and one algorithm over all the others, just on the basis of similarity. Second, and more importantly, not all similarities are of equal significance for species groupings. Some small genetic differences can have disproportionately large effects, and have disproportionate significance relative to evolution and development. Third, some differences, such as sexual dimorphism and developmental stages, don't seem appropriate for determining species membership. Any way of dividing species that places reproducing organisms into different species based on sex alone seems problematic. So while similarity may be relevant to species groupings it cannot be determinative by itself.

The third main type of species concepts is based on the idea that species are historical entities. The have beginnings, endure over time, and eventually end in speciation or extinction. This is perhaps most apparent in the fossil record but it is also implicit in the evolutionary tree in Darwin's *Origin*, and in the tree thinking just discussed in Chapter 5. The *evolutionary species concept*, advocated by Mayden and E. O. Wiley and based on the views of G. G. Simpson, asserts that a species is "an entity composed of organisms which maintains its own independent evolutionary fate and historical tendencies" (Mayden 1997, 395; Wiley and Mayden 2000). Other historical concepts include the *successional species concept*, used for identifying fossil taxa and based on the idea that change inferred in a lineage from fossil remains can be subdivided into multiple successive species. Similarly, the *paleospecies concept* and the *chronospecies concept* each conceive of species as segments of a changing lineage (Mayden 1997, 410–411).

Some historical concepts focus on branching speciation rather than change over time. These include the *cladistic species concept*, the *composite species concept*, the *intermodal species concept*, and the *phylogenetic species concept* (Mayden 1997). On these approaches organisms are part of a species if they are part of a segment of a lineage that originated in a speciation event. This way of thinking about species seems to be a natural consequence of the tree thinking we discussed in the last chapter. The specifics of each of these concepts are not important for purposes here, but what is important is the fact that species taxa, on each of these historical concepts, are groups of organisms that are linked together *historically* in a pattern of ancestry and descent.

So what do we do with all of these inconsistent ways of thinking about species – this use of conflicting species concepts? We cannot adopt all of them because they conflict in application, dividing organisms into species differently. The turn to a *phylogenetic species concept* from other concepts, for instance, has in one case multiplied 15 amphibian species into 140 (MacLauren and Sterelney 2008, 28). A recent survey of research quantifies the effects of a shift to this particular species concept from other concepts, finding a 300 percent increase in fungus species, a 259 percent increase in lichen species, a 146 percent increase in plant species, a 137 increase among reptile species, an 88 percent increase among bird species, an 87 percent increase among mammals, a 77 percent increase among arthropods, and a 50 percent decrease in mollusc species (Agapow et al. 2004, 168).

This use of different species concepts that divide biodiversity differently seems to lead to two kinds of pluralism: a *category* pluralism, in the different ways to think about the species rank; and a *taxa* pluralism in how organisms are grouped and divided into species. These pluralisms are related in that those who use different species concepts typically group organisms into species taxa in different ways. Category pluralism typically leads to taxa pluralism. Notice though, that even if two researchers adopt the same species concept – they are category monists, they might still group organisms into species differently, based on different similarity algorithms, individual grouping and splitting tendencies, and more.

What are the implications of these pluralisms? First, it isn't clear that they automatically threaten the whole project of biological classification. There could be uncertainty about the species category and species taxa without uncertainty about the genus and higher-level categories. Different species grouping schemes *might* be compatible with a single genus level grouping in the sense that even on the different ways of grouping, all of the species remain part of the same genus. So the evolutionary tree may have some limited uncertainty at the species tips of its branches without calling into question the overall structure of the tree. Radical differences in species groupings would pose a greater risk, however, if organisms were grouped by conflicting species concepts such that they would be members of different higher-level categories, and would therefore be on different branches of the evolutionary tree.

But the category and taxa pluralisms surely put the species level of classification at risk, along with the assumptions that species are the units of evolution and biodiversity. Most serious is an antirealism worry based on taxa pluralism: There are many ways to group organisms into species, based on different species concepts, because there really are no species things out there in the world. On this antirealism, our divisions of organisms into species taxa are just useful ways to think about nature. Perhaps less serious is the worry based on category pluralism. There is no single *kind* of species thing, and so different species taxa are not comparable. This pluralism leaves open the question as to whether species are units of evolution or biodiversity. Some species taxa may be, some not. Different kinds of species taxa need not all play the same roles relative to biodiversity and evolution.

Varieties of Pluralism

Implicit in the discussions about species concepts are three main evaluative criteria. First, species concepts are typically evaluated in terms of *operationality*: How easy are they to apply in grouping and dividing organisms into species? Species concepts that can be easily applied, such as those based on morphological similarity, are typically regarded as better on these grounds than those not easily applied. Second is *theoretical significance*: How appropriate is a species concept relative to the demands and commitments of evolutionary theory? Since species are often taken to be evolutionary units, then one concept can be better than another insofar as it better reflects the evolutionary roles of species. Criteria based on reproductive isolation and cohesion, for instance, reflect the idea that species taxa are formed and maintained in part through these processes. Perhaps the most obvious criterion for the evaluation of species concept is *universality*: A species concept is better insofar as it applies across biodiversity. The primary criticism of the *biological species concept*, for instance, is that it applies only to sexually reproducing organisms.

An ideal species concept would then apply to all organisms across biodiversity, be operational, and have the right theoretical significance (Hull 1997, 357–380). If there were such an ideal concept, then the grouping of organisms into species taxa would be relatively unproblematic. A universal concept could group all organisms in a theoretically correct way, and would do so with relative ease based on its operationality. If the theoretical basis were correct, then we could be confident we were "cutting nature at its joins." But as the preceding survey of species concepts suggests, no single species concept seems ideal. The most theoretically relevant concepts are typically neither operational nor universal, and the most operational and universal concepts are typically not theoretically relevant. In short, the reason we still have the species problem is that no single concept seems to satisfy all our criteria.

One obvious response to this problem is to be a pluralist about the species category, embracing the fact that no single species concept seems adequate, and that there must therefore be an irreducible plurality of species concepts. And one way to be a pluralist is on pragmatic grounds. On this approach, the reason there are multiple species concepts is that there is no single right way to divide and group organisms in all contexts. Rather there are many ways to do so depending on our theoretical interests. We might be interested in morphology, and therefore could group on the basis of morphological similarities and differences. Or, we might be interested in genetic differences and similarities. We could then group on those grounds. Or we might be interested in the lineages hinted at in the fossil record. If so, we could think of species in terms of lineages. Or, we might be interested in the processes that are responsible for the production and maintenance of reproductive isolation. If so, we could think of species in terms of evolutionary processes. On this approach, there is no *single* correct way to think about species; rather there are many legitimate ways to think about them. Philip Kitcher has argued for such a pragmatic pluralism. According to Kitcher:

> Species are sets of organisms related to one another by complicated, biological interesting relations. There are many such relations which could be used to delimit species taxa. However, there is no unique relation which is privileged in that the species taxa it generates will answer to the needs of all biologists and will be applicable to all groups of organisms. (Kitcher 1992, 317)

There are, according to Kitcher, multiple ways to conceive species because there are multiple kinds of biological investigations, based on different theoretical interests. (See also Dupré 1993, 2012; Slater 2013.)

The advantage of this *pragmatic pluralism* is obvious. It makes sense of the many different ways biologists think about species in pursuing their own research, from morphology, genetics, and interbreeding to ecological functioning. Researchers tend to focus on what is central for them and their particular research agenda, as Kevin de Queiroz explains:

> The existence of diverse species concepts is not altogether unexpected, because concepts are based on properties that are of the greatest interest to subgroups of biologists. For example, biologists who study hybrid zones tend to emphasize reproductive barriers, whereas systematists tend to emphasize diagnosability and monophyly, and ecologists tend to emphasize niche differences. Paleontologists and museum taxonomists tend to emphasize morphological differences, and population geneticists and molecular systematists tend to emphasize genetic ones. (de Queiroz 2005, 6601)

So if what we are interested in is *explaining the behavior* of those who use species concepts, pragmatic pluralism certainly succeeds.

The problem with this approach though is obvious as well. What counts as a species depends on the theoretical interests and needs of the investigators. So as theoretical interests diverge, species concepts diverge, and species counts diverge. This category pluralism seems to lead to a taxa pluralism – multiple ways of dividing and grouping organisms into species. But this way of identifying and individuating species also seems subjective in that it depends on the contingent theoretical interests of researchers. One researcher with her interests will group one way, and another researcher with his interests will group another. In fact, a single researcher could have different theoretical interests at different times, one day adopting one species concept and grouping organisms one way, the next day adopting a different species concept and grouping organisms differently. But what if a researcher also has the interest in constructing a *single* unambiguous classification? That research interest does not even seem to be satisfiable by this pragmatic pluralism. Moreover, how would one apply endangered species legislation? To do so, a single preferred species grouping and concept would be required. Pragmatic pluralism seems to suggest that the solution here is political – pick the favored theoretical interests and species groupings of some researchers over the theoretical interests and species groupings of others. Similarly, if we were to maintain that species were the units of evolution then to determine what the units of evolution were, we would need to decide whose research interests we prefer. But surely we shouldn't be deciding theoretical issues in science on this seemingly political and subjective basis. The bottom line is this: If there are no good, nonsubjective reasons to prefer the research interests of one group over another, then preferences for species concepts would seem arbitrary. But if there is a nonsubjective reason to prefer the research interests of some researchers, why not do that from the beginning? Talk about research interests would then be dispensable.

Kitcher hasn't seemed concerned about this subjectivity, dubbing his approach "pluralistic realism," based on the idea that the groupings of the various researchers are based on real features of the world. The similarities and differences, processes, or historical relations are, after all, real. To return to Plato's metaphor: We can "cut nature at its joints" in different ways – at different joints. But still it is hard to see how the groupings themselves can be real in a nonconventional *natural* sense, if they are ultimately dependent on subjective factors such as the contingent interests of researchers. To extend the butcher metaphor: One can cut at different joints, but one cannot cut the *same parts* at different joints.

In the first chapter we contrasted conventional and natural kinds on the grounds that the conventional are dependent on human preferences and practices. Pragmatic species grouping could certainly be real in this conventional sense, just as dollar bills are real and universities are real. And researchers could come to agree on these conventional ways of grouping. But the insight from the distinction between natural and conventional kinds is that some groupings are independent of human practices and preferences and this seems to be a significant difference for science. Surely our attitudes toward species groupings would be different if they were *merely* conventional. The preservation of conventional groupings, based on researchers' interests, hardly seems to have the same value and significance as the preservation of natural groups that are not so based.

Other forms of pluralism do not seem to challenge species realism to the same degree. *Ontological pluralism* is based on the idea that there are different species concepts because there are different kinds of species things in nature. In other words, the term 'species' is systematically ambiguous, referring to different kinds of things. On this way of thinking, big vertebrate species might be different from insect, plant, and bacterial species, even though each species grouping cuts nature at its joints. Most obviously, sexually reproducing and asexually reproducing species are different kinds of things, based on the cohesion in the former through sexual reproduction, and lacking in the latter.

Marc Ereshefsky advocates such an ontological pluralism, arguing that we should conceive species as historical lineages that are part of the evolutionary tree, but there are different kinds of lineages based on the different kinds of evolutionary forces that operate within each lineage. He claims that there are three main kinds of species lineages based on three processes: interbreeding, ecological, and monophyletic.

> All of the organisms on this planet belong to a single genealogical tree. The forces of evolution segment that tree into a number of different types of lineages, often causing the same organisms to belong to more than one type of lineage. The evolutionary forces at work here include interbreeding, selection, genetic homeostasis, common descent, and developmental canalization ... The resultant lineages include lineages that form interbreeding units, lineages that form ecological units, and lineages that form monophyletic taxa. (Ereshefsky 2001, 139)

These different kinds of lineage concepts apply in different ways to biodiversity. Some lineages will form interbreeding units; others will not. Some will form ecological units; others will not.

But couldn't there be a yet more fundamental and unifying species concept? After all, if species are lineages, then surely the idea of a lineage is more fundamental. Ereshefsky rejects this possibility because the lineage requirement by itself cannot distinguish species taxa from higher-level taxa, which are also lineages:

> So, is there a common unifying feature of species taxa? Each species taxon is a genealogical entity, but that commonality is too inclusive - it includes all taxa. Species taxa are maintained by different processes, so they lack a common type of unifying process. Finally, many phylogenetic and interbreeding species fail to be cohesive or unitary in the same way. What is left as the common feature of species taxa is the term 'species', and a widespread motivation among biologists to find the base taxonomic and evolutionary units of the Linnaean hierarchy. (Ereshefsky 1998, 113)

Hence, according to Ereshefsky, we are left with an irreducible pluralism. There is, however, a possibility not considered here by Ereshefsky. It is not that species taxa are lineages, but they are *segments* of lineages. This is made most clear by how species are typically represented on the evolutionary tree. Species on the tree are not entire branches, with many sub-branches and twigs. Rather they are *sections* of branches between nodes or branching points. Species begin in a branching event, and then end in a branching event, or at the crown of the tree. Higher-level taxa may originate in a speciation event, but unlike species, they also persist over speciation events. If we look at the evolutionary tree, the difference between species lineages and higher-level lineages is clear.

One advantage of this ontological pluralism is that it explicitly recognizes that there are important differences across biodiversity. What might be responsible for speciation among sexually reproducing vertebrates is not necessarily found among invertebrates, insects, bacteria, fungi, viruses, and so forth. But an implication of ontological pluralism is that if there are different kinds of species things, then not all species taxa are equivalent. Bacterial species may not serve as units of evolution and biodiversity to the same degree that vertebrate species do. Moreover, we may not regard all species taxa as being "real" to the same degree. (We will return to this idea in Chapter 7.)

One disadvantage of Ereshefsky's ontological pluralism (as suggested in the preceding passage) is that it seemingly allows for the cross classification of individual organisms into different species and different kinds of species. An organism may potentially be part of two different kinds of lineages, and hence members of two different species – and two different *kinds* of species. There will also likely be a difference in species counts as Ereshefsky admits: "…when surveying a plot of land we might find that the interbreeding approach identifies three distinct species whereas the phylogenetic approach identifies two species…" (Ereshefsky 2001, 156). But here Ereshefsky seems to interpret his own approach pragmatically, suggesting that we can simply adopt different approaches. Unless an ontological pluralism can generate unequivocal classifications, it will lead to the same taxa pluralism and conflict in species counts we saw with the pragmatic pluralism. This kind of ontological pluralism seems to lead to pragmatic pluralism.

Another variety of ontological pluralism is a *ranking pluralism*. Brent Mishler and Michael Donoghue advocate such an approach. (See also Mishler and Brandon 1987.) First, they adopt the general cladistic view that species, like all biological categories, are monophyletic groupings – groups consisting only of an ancestor and all its descendants. This is a universal *grouping* criterion. But there is no single criterion that distinguishes the species level of monophyly from the other levels:

> Even when monophyletic groups are delimited, the problem of ranking remains since monophyletic groups can be found at many levels within a clade. Species ranking criteria could include group size, gap size, geological age, ecological or geographic criteria, degree of intersterility, tradition, and possibly others. The general problem of ranking is presently unresolved, and we suspect that an absolute and universally applicable criterion may never be found, and that, instead, answers will have to be developed on a group by group basis. (Mishler and Donoghue 1992, 131)

Since the ranking of monophyletic groups into species is done on the basis of multiple criteria, there must be multiple species concepts: "...we think that a variety of species concepts are necessary to capture the complexity of variation patterns in nature" Mishler and Donoghue claim that this is an unequivocal species concept in that there is a single *grouping* criterion – monophyly, even if there are multiple ranks of monophyletic groups, and *ranking* is determined by multiple criteria (Mishler and Donoghue 1992, 131).

The obvious problem for this approach is that it isn't obvious how to apply the idea of monophyly at the species level. Usually, and as we saw in Chapter 4, monophyly is defined in terms of *species*, not individuals: A group is monophyletic if and only if it contains an ancestral species, and all and only the descendant species. But this definition could not possibly apply at the species level itself. Species are not composed of other species, but of organisms. Mishler and Donoghue recognize this is a problem, but don't offer a solution:

> Several different concepts of monophyly have been employed by systematists, but none explicitly at the species level ... We favor Hennig's concept of monophyly (except explicitly applied at the species level) but are fully aware of the difficulties in its application at low taxonomic levels ... In particular the difficulty posed by reticulation. (Mishler and Donoghue 1992, 130)

As suggested at the end of this passage, within biparental species and at the species level there is reticulation through sexual reproduction. If so, there can be no single ancestral organism of the members of a species. At minimum there would be an ancestral couple. Moreover, the offspring of this ancestral couple would presumably be interbreeding with other individuals of different ancestry. More plausibly we should think of the ancestral organisms as a *population*. The bottom line is that it isn't at all clear how to apply the idea of monophyly to species – or that it even could be applied.

There is yet another version of pluralism, a *hierarchical pluralism*, based on a division of conceptual labor, and developed by Richard Mayden (1997) and Kevin de Queiroz (1999), and advocated in Richards (2010). This approach begins by noting that some species concepts serve a theoretical function, telling us what kinds of things species are. Other concepts seem to serve an operational function, telling us how to identify and individuate species. This approach thereby rejects the view that species concepts are *each* to be evaluated in terms of universality, operationality, and theoretical significance. Rather some species concepts are to be evaluated on operationality, and others on universality or on theoretical significance. If there is this division of conceptual labor, then the question becomes: Which species concepts should we treat as theoretical? According to Mayden, theoretical concepts should be judged on theoretical significance and universality:

> What then are the criteria we should be looking for in a primary concept? It should be consistent with current theoretical and empirical knowledge of diversification. It should be consistent with the ontological status of those entities participating in descent and other natural properties ... Finally, it should be general enough to encapsulate all types of biological entities considered species as taxa by researchers working with supraspecific taxa. (Mayden 1997, 419)

Since theoretical concepts are to be judged on the basis of theoretical significance, we can begin with what evolutionary theory tells us about species. So what does evolutionary theory tell us about species? First it tells us that species exist over time – *diachronically*. They have origins in speciation, duration in which there might be substantial change, and endings in extinction or new speciation. Second it tells us that species exist at a single time – *synchronically*, as populations or groups of populations, that vary, and that sometimes have cohesion through sexual reproduction, social interaction, common ecological functioning, and selection regimes. A satisfactory theoretical concept presumably must reflect these theoretical commitments. A primary theoretical concept should also be universal so as to apply to all things that evolutionary theory tells us are species.

Mayden argues that the best candidate for primary theoretical concept is the *evolutionary species concept (ESC)*, proposed by G. G. Simpson, which asserts that a species is "a lineage (an ancestral-descendent sequence of populations) evolving separately from others and with its own unitary evolutionary role and tendencies" (Mayden 1997, 395). This is a historical species concept in that it focuses on the temporal, diachronic dimension of species. Species are things that have beginnings, perhaps change over time and an ending. This is also a population based, synchronic concept in that the lineages exist at a single time as a population or group of populations. Because this population lineage is evolving separately from others and with its own role and tendencies, there must be some sort of cohesion, whether through reproduction, social interaction, selection pressures, or some other process. The advantage of this way of thinking about species, according to Mayden, is that it can be universal. In principle it applies, for instance, to both sexually and nonsexually reproducing organisms. *Whatever* processes are at work in the origin and maintenance of species, species are all ancestral-descendent sequences of populations.

Similarly, de Queiroz argues for an approach based on the same basic idea, the *general lineage concept*:

> Species are segments of population-level lineages. This definition describes a very general conceptualization of the species category in that it explains the basic nature of species without specifying either the causal processes responsible for their existence or the operational criteria used to recognize them in practice. It is this deliberate agnosticism with regard to causal processes and operational criteria that allows the concepts of species just described to encompass virtually all modern views on species, and for this reason, I have called it the general lineage concept of species. (de Queiroz 1999, 53)

As de Queiroz makes clear, this *general lineage concept*, like the *evolutionary species concept*, is potentially universal because it is vague about which evolutionary processes are relevant to the formation and maintenance of a particular species.

According to both Mayden and de Queiroz, the primary theoretical species concept has the disadvantage in that it is not operational; it cannot be applied directly to observation. After all, one cannot just look at an organism or group of organisms and see that it is part of a segment of population lineage that is subject to particular causal processes in the same way one can look at an organism and see its morphological similarity to other organisms. Some species concepts are operational in the sense that they can be easily applied to group and divide organisms into species - given a theoretical concept. These "secondary" concepts will also be required, according to Mayden:

> While the ESC is the most appropriate primary concept, it requires bridging concepts permitting us to recognize entities compatible with its intentions. To implement fully the ESC we must supplement it with more operational, accessory notions of biological diversity - secondary concepts. Secondary concepts include most of the other species concepts. While these concepts are varied in their operational nature, they are demonstrably less applicable than the ESC because of their dictatorial restrictions on the types of diversity that can be recognized, or even evolve. (Mayden 1997, 419)

De Queiroz recognizes this same hierarchy of species concepts, but uses the term 'criteria' to refer to Mayden's operational concepts.

> The species criteria adopted by contemporary biologists are diverse and exhibit complex relationships to one another (i.e. they are not necessarily mutually exclusive). Some of the better-known criteria are: potential inter-breeding or its converse, intrinsic reproductive isolation ... common fertilization or specific mate recognition systems ... occupation of a unique niche or adaptive zone ... potential for phenotypic cohesion ... monophyly as evidenced by fixed apomorphies ... or the exclusivity of genic coalescence ... qualitative ... or quantitative ... Because the entities satisfying these various criteria do not exhibit exact correspondence, authors who adopt different species criteria also recognize different species taxa. (de Queiroz 1999, 60)

For both Mayden and de Queiroz, these operational criteria should *not* be judged in terms of universality because they pick out processes and features that are specific to just some species taxa, and within a limited range of biodiversity. The sexual reproduction criteria, for instance, are appropriate for sexually reproducing organisms, but not for those that reproduce asexually. And a criterion based on morphological similarity will be appropriate for those organisms in population lineages that share the right kind of similarities – but only for the appropriate patterns of similarity. Strong sexual dimorphism, for instance, need not result in different species groupings of males and females because morphology is relevant only in specified ways.

If this hierarchical pluralism is right, then the species problem is a result of the confusion of different kinds of species concepts – those that tell us what species are, and those that tell us how to group and divide organisms into species. The solution to the species problem is then a better understanding of the conceptual framework. Perhaps we should then change our terminology to reflect the fact that operational species concepts function differently than theoretical. De Queiroz calls these operational concepts "criteria." Or we can follow the lead of Rudolf Carnap, in a debate from physics early in the twentieth century, whose "correspondence rules" connect theoretical principles to observation (Richards 2010, 139–141). If so, then the operational concepts are really correspondence rules for connecting a theoretical species concept to observation.

But whether or not we adopt this alternate terminology, we can see some of the advantages of this *hierarchical pluralism*. First, it explains the use of different concepts by researchers. Which operational criteria are relevant depends on which organisms are in question, and the particular evolutionary processes that are relevant to the origin and maintenance of species in these organisms. And since there might be multiple processes that are relevant, use of a criterion might also depend on which of these processes are of one's own research interests. So two researchers could study the same set of organisms, but use different criteria – different correspondence rules, based on different training and current interests. A geneticist and morphologist could study the same group of organisms but think about them in terms of genetic concordance or morphological similarity. Similarly, an ecologist and physiologist could think about a group of organisms in terms of the functioning of natural selection or mate recognition systems. De Queiroz makes this case:

> Although all modern biologists equate species with segments of population lineages, their interests are diverse. Consequently, they differ with regard to the properties of lineage segments that they consider most important, which is reflected in their preferences concerning species criteria. Not

> surprisingly, the properties that different biologists consider most important are related to their areas of study. Thus, ecologists tend to emphasize niches; systematists tend to emphasize distinguishability and phyly; and population geneticists tend to emphasize gene pools and the processes that affect them. Paleontologists tend to emphasize the temporal extent of species, whereas neontologists tend to emphasize the segments of species that exist in the present. (de Queiroz 1999, 65)

But because the different researchers still adopt the same theoretical concept, they can use different *criteria* without generating conflicting groupings. The important insight here is that these different ways of thinking about species are not conflicting ways of thinking, but complementary – as long as it is not assumed that each criterion must be universal. And the more we know about organisms and the evolutionary forces that operate on them, the more criteria we will have for grouping and dividing them into species. With these secondary operational concepts, criteria, or correspondence rules, the more the merrier! Here multiplicity rather than universality is our goal.

Notice the difference between the pragmatic and hierarchical pluralisms. Both can explain the differences in emphasis among researchers. It is legitimate under both for morphologists to focus on morphological differences and similarities, for those interested in processes of speciation to focus on reproductive isolation, and for those interested in paleontology to focus on the patterns in the fossil record. But for hierarchical pluralism, the single theoretical concept also constrains research interests. Morphologists are free to focus on some morphological differences as indicators of species membership, but not on others, such as sexual dimorphism or developmental stages. The theoretical concepts advocated by de Queiroz and Mayden tell us that these differences are irrelevant to the identification and individuation of species as segments of population lineages. By contrast, on the purely pragmatic approach it isn't clear how researchers' interest can be theoretically constrained. Any one researcher's interest is seemingly just as legitimate as the interests of any other researcher.

In fact, we could take this pragmatic pluralism even further and argue that any classification is good if it serves the research interests of anyone, whether or not he or she is a biologist. A physicist, for instance, might have a legitimate research interest in a group of organisms based on ratio of surface area to mass. But the grouping that serves the physicist's research interests does not even plausibly coincide with what we might typically think of as species. In other words, this grouping based on research interest is certainly not a *species* grouping. Moreover, it is not clear that the research interests of biologists always coincide with what are plausibly *species* groupings. This points to a problem with pragmatic pluralism. There is no reason to believe that the groupings that serve any particular research interest are what we might plausibly claim to be *species* groupings. Perhaps we could constrain research interests by the intentions of researchers: Only the research interests of those who are intending their research to group at the species level would be relevant in determining species groupings. But this seems too subjective. Why would the mere intention to produce a species grouping be sufficient for determining species grouping? Moreover, we would have the problem of determining precisely when a researcher has that particular intention. This is surely a problem for pragmatic pluralism if it leads us down this path.

The Problem with Microbes

Hierarchical pluralism assumes, as indicated in the first sentence of the de Queiroz quote given earlier, that there is universal agreement about the primary theoretical concept. It is difficult to know if this is literally true, but it is hard to see how anyone could accept modern evolutionary theory and deny that species taxa have the basic features of segments of population lineages in that they extend over space and time, and have beginnings and endings. But the insight of *ontological pluralism* is that there are important differences among species across biodiversity. Perhaps we should explicitly recognize these differences in how we conceive species. If so, no single theoretical species concept will do because there are significant differences among the population lineages that should be reflected in our theoretical species concepts.

This worry is sharpest relative to microorganisms such as prokaryotes and viruses. One of the fundamental divisions in biodiversity is between those organisms that have a membrane-defined nucleus that holds the genetic material - the eukaryotes - and those that do not - the prokaryotes. Eukaryotes mostly reproduce sexually, and can maintain reproductive isolation through differential mating - mating only with other individuals within a single population lineage. One exception among eukaryotes, as we saw in Chapter 5, is the hybridizing plants that form new species by reproduction across species boundaries. The number of plant species that hybridize or are formed through hybridization is a minority, but still significant. In one survey of five flora from Europe, North America, and Hawaii, the authors report that from 6 to 16 percent of the genera contain at least one hybrid (Ellstrand, Whitkus, and Reiseberg 1996, 5090). The question here is this: Are these plants in which hybridization has occurred different enough from the nonhybridizing organisms we see among most plant and animal species to justify treating them as different kinds of species things? In other words, do we need an additional theoretical concept for hybridizing plants?

The problem is even more acute when we look at the prokaryotes – the Bacteria and Archaea that reproduce asexually, but also share genetic material through lateral or horizontal gene transfer (HGT). As we saw in Chapter 5, HGT is the result of several processes: *transduction*, where genetic material is transmitted from one bacterium to another by viral bacteriophages; *conjugation*, where genetic material is transmitted through direct contact; and *transformation*, the uptake and incorporation of exogenous genetic material. This produces reticulation in the evolutionary tree through the transfer of genetic material across population lineages. Are the differences between the prokaryotes and sexually reproducing eukaryotes such that we should treat them as different kinds of species? And from the approach based on hierarchical pluralism is a more specific question: Are the differences between the *population lineages* of the prokaryotes and sexually reproducing eukaryotes such that we should treat them as different kinds of lineages?

This same problem with HGT is found to an even greater degree among the "bacteria eating" viruses, the bacteriophages. Bacteriophages, perhaps the most numerous biological agents on the planet, are responsible for some of the HGT among bacteria, but they also exchange genetic material at a higher rate among themselves. HGT can occur if two or more bacteriophages infect a single bacterium, each releasing genetic material that can be recombined. This is relatively common, and some researchers think that most bacteriophages are *mosaic*, containing genetic material from multiple different kinds of bacteriophages (Morgan and Pitts 2008, 748). Consequently, a genetic criterion will be problematic. If a given bacteriophage contains strings of genetic material acquired through HGT from multiple different bacteriophages, it isn't clear which string of genetic material should be taken as determinative for classification.

Notice a phenetic criterion would be problematic for the same reasons. Unlike in cases with predominantly vertical genetic transmission, such as we see in the sexually reproducing organisms, the different phenetic features in viruses, such as tail and head morphology, do not classify concordantly. A single bacteriophage might have tail morphology from one bacteriophage source and head morphology from another. If we classify on the basis of tail morphology, we get one species grouping. If we classify on the basis of head morphology we get another species grouping. In spite of these problems, the International Committee on the Taxonomy of Viruses has attempted to assign every virus to a set of Linnaean categories – species, genus, family, and order (Morgan and Pitts 2008, 746).

Perhaps this mosaicism threatens the idea not just that viruses can be represented on the evolutionary tree (as discussed in Chapter 5), but that they even form species at all. Gregory Morgan and Brad Pitts have recently claimed that bacteriophages don't form species on any of the main species concepts. The biological species concept based on sexual reproduction, for instance, clearly cannot apply to bacteriophages, because they don't reproduce sexually in the usual sense (Morgan and Pitts 2008, 753). Morgan and Pitts also reject concepts based on phenetic and genetic similarity because different phenetic features and genetic sequences cross classify. One feature and its genetic sequence in a bacteriophage could come by HGT from one source and another from a different source. They also claim that an ecologically based species concept fails in part because it is "difficult to define what an ecological *niche* is for a virus," and "...one would be challenged to find a terrestrial extracellular environment that they do *not* occupy" (Morgan and Pitts 2008, 744–745). We could not, in other words, group bacteriophages into species based on the bacteria that can infect because bacteriophages can in general infect many different kinds of bacteria.

A phylogenetic species concept also has problems, according to Morgan and Pitts, as it identifies species with the segmented branches of the evolutionary tree. An individual organism is a member of a species if and only if it is part of the lineage of a particular branch. But it is not clear how this way of thinking can apply to highly mosaic bacteriophages. First, with each case of HGT, there will be a fusion of different bacteriophages. Will the offspring of this fusion be a new species (Morgan and Pitts 2008, 754)? If so, then there would likely be as many bacteriophage species as instances of HGT between very different bacteriophages. The second problem is a consequence of the first. Because of HGT, each bacteriophage will in general have many different origins depending on the mosaic sources of genetic material. If so, there wouldn't be any *single* lineage to determine species membership. But suppose there is a species associated with each lineage. If so, then a single bacteriophage would be a member of *many* species taxa, since it has genes from many lineages. The dilemma here is that each bacteriophage might either be a single species itself, or a member of many different species lineages.

Ultimately, the problem is that it isn't clear that there are any such things as discrete lineages among bacteriophages at all. Rather there are just many interconnections within a large and complex network. This is the problem we saw with tree thinking relative to bacteria, but is found to an even greater degree with bacteriophages. This brings us back to the question posed at the beginning of this section about whether a single theoretical species concept, such as one based on de Queiroz's *general lineage concept* or Mayden's *evolutionary species concept*, can apply across biodiversity. Is it possible for a single theoretical species concept based on the general idea of a population lineage to apply to eukaryotes, prokaryotes, and viruses? Even if the idea of a population lineage is relatively clear for sexually reproducing vertebrates and can plausibly serve as a foundation for thinking about species within specific domains, it isn't clear that this idea can be applied to hybridizing plants, prokaryotes, and viruses. In cases of hybridization and massive HGT, it is not clear there will be well-defined boundaries to a population lineage.

So how should we respond to this challenge posed by HGT? One possibility is to just deny that there are species in organisms where HGT is common, as in bacteriophages. This is how Morgan and Pitts respond:

> In bacteriophage evolution, adaptation to changing environmental conditions – mostly due to the population dynamics of the host bacteria – and prolific HGT results in individuals that are composed of convoluted chimerical genomes and phenotypes that reflect no one particular evolutionary history. There is good reason to think that the Earth's most abundant biological agent evolves without good and distinct species. (Morgan and Pitts 2008, 762)

Ereshefsky similarly denies that prokaryotes form species, arguing that none of the three kinds of species lineage he recognizes – interbreeding units, ecological units, and monophyletic taxa – can apply. He then considers another possibility, that prokaryotes form *composite evolutionary units*, which are "integrated associations of lower-level elements replicated and held together by biological mechanisms." These units are composite because they consist of "phylogenetically diverse genes" (Ereshefsky 2010, 560). Perhaps bacteriophages and prokaryotes should not be conceived in terms of the Linnaean framework at all, but within an entirely different framework based on the generic idea of composite evolutionary units. .

There is another way to think here, based on the idea that specieshood comes in degrees. Some organisms form more cohesive, clearly defined and distinct species, while others form species that are less so. In the framework of the hierarchical pluralism this would be understood in terms of population lineages broadly construed. Some species form more cohesive, highly defined and distinct population lineages, others less so, and with gradations in between. So we might have at one end of this continuum the sexually reproducing vertebrate species that have strong reproductive isolation. And at the other end we might have the bacteria and bacteriophages that form loose, ill-defined, and indistinct population lineages that have substantial reticulation. We can understand this idea by analogy with organisms. Some organisms, such as we find among vertebrates, have strong cohesion, with organ systems causally interacting at a variety of levels. At the other end we find the slime molds and organisms that often seem to be more like colonies. These have less tightly integrated organisms are therefore less real in some sense. We will return to this idea in Chapter 7.

Conclusion: Are Species Real?

So where are we now with the species problem – the use of multiple inconsistent species concepts? According to pragmatic pluralism, there are multiple kinds of species things based on research interests. If so, then there will be an irreducible taxa pluralism – multiple, inconsistent ways to divide organisms into species with no obvious resolution. This seems to threaten the idea that species are real things that exist in the world independent of human conventions or preferences. This also threatens our objectivity in theorizing about species and in the conservation of species taxa. My species may not be your species, which may not be anyone else's species. An ontological pluralism might avoid this worry if there were no possibility for taxa pluralism – classifying organisms into species in different and inconsistent ways. But as we have seen, Ereshefsky's pluralism doesn't avoid this problem. Perhaps an as yet undeveloped ontological pluralism could do so. The hierarchical pluralism of Mayden and de Queiroz looks more promising as a solution. Because it is monist in that there is a single theoretical concept, there is at least the possibility that we can reach agreement about what counts as a species, insofar as we can divide biodiversity into the right kind of segments of meta-population lineages. But even so, it isn't clear that we can think about the full range of biodiversity, from vertebrates to bacteriophages, in terms of segments of population lineages.

There are several possible outcomes here. First, the species problem never gets resolved. We just more or less ignore or work around the problems it poses for species counts, endangered species legislation, classification, and evolutionary theory. Given the fact that the species problem already has a long history, this might be the most probable outcome. Second, the species problem might also get resolved over time based on purely philosophical arguments. But given the long endurance of many philosophical disputes, such as those over free will and consciousness, this also seems unlikely. The third possibility is that the species problem will get resolved in terms of the theoretical demands and practices of biology. Perhaps this is already happening, if De Queiroz is right in a passage quoted earlier in this chapter: "all modern biologists equate species with segments of population lineages" (de Queiroz 1999, 65). There *might be* an implicit acceptance of a particular view about species, without an explicit recognition of a single theoretical concept. If so, then there is already some sort of resolution to this dispute. But then what do we do with all of the discussion about species that seems to instead be committed to a morphological, genetic, or interbreeding concept?

Part of understanding the species problem may be found in how language works. In *The Species Problem* (2010), I argue that the term 'species,' like many theoretical terms, has a structure – a definitional core and descriptive periphery. We can understand this structure in terms of descriptive content. The definitional core has a descriptive content that counts as a definition. The descriptive periphery also has descriptive content, based on other features, traits, or properties associated with the referent of the term (what it applies to) without being definitional. For the term 'water,' for instance, there is a definitional core consisting in the descriptive content, in terms of its molecular composition of one oxygen and two hydrogen atoms. Water is by definition H_2O. This is what we take to make something water, so it counts as a definition. But there is also a great deal of descriptive periphery here as well. Water is wet, good for quenching thirst, used in swimming pools, covers most of the Earth's surface, falls as rain and snow, is used to cool nuclear reactors, and so on. These are facts about water, but are not constitutive in that they don't define the term 'water.' We can distinguish the definitional core and descriptive periphery for theoretical terms in general. 'Gold' may be defined by its atomic number – 79, but it also has many other contingent features – a particular price on the market, a use in jewelry, a particular color, density, and so on. In both cases, the definitional core tells us what counts as a certain kind of thing, while the descriptive periphery includes a lot of other nondefinitional content (Richards 2010, 178–204).

This understanding of theoretical terms and the associated concepts can help us understand the linguistic practice associated with 'species.' Evolutionary theorists can focus on the definitional core, but other researchers can focus on whatever part of the descriptive periphery is in their theoretical interests. So geneticists can focus on the genotypes associated with a particular species, while population geneticists can focus on the frequency and distribution of particular genes within and across population. Similarly, morphologists can focus on morphology, paleobiologists on changes in an archaic lineage over time, behaviorists on mate recognition systems, and so on. But these interests in the descriptive periphery are no more constitutive of the theoretical species concept than a jeweler's interest in the ductility of gold would be constitutive of the theoretical concept of gold. If this picture is right, part of the origin of the species problem is in a misunderstanding of how language works in science (Richards 2010, 202–204). The solution to the species problem would then consist partly in the clarification of language. If so, the species problem is not so much solved as dissolved.

This linguistic solution doesn't solve all of the problems though, even if its analysis were basically correct. Chemists didn't discover the atomic number of gold or molecular structure of water through linguistic analysis. It happened through the practice of chemistry, and the theoretical debates about elements and compounds. This is where the hard work occurred. Similarly, the hard work here, relative to the species problem, will be in the theoretical and empirical debates about specific species taxa and the species category. How important are species to classification and evolutionary theory? What sorts of things must they be if our theories are right? Does all of biodiversity come packaged in species taxa? Is there a single kind of species? How all this will work out is not yet clear. And although I think it is unlikely, we may even decide that not only is there no species category, there really are no such things as species taxa. This possibility will, however, have a high price if we want to continue treating species as units of biodiversity, classification, and evolution.

7 The Metaphysics of Biological Classification

Metaphysics

In Chapter 6 we examined the species problem - the use of multiple, inconsistent species concepts. Some biologists think about species as interbreeding populations and the processes that produce new species and maintain those species. Others think about them in terms of patterns of similarity and difference in morphology, genes, and behavior. Yet others think about them in terms of lineages, and as branches on the evolutionary tree. The problem with all of these ways of thinking about species is that no single way of thinking seems adequate and the many ways of thinking seem inconsistent with each other.

While there is no consensus yet about any of the proposed responses to this problem discussed in Chapter 6, perhaps there has been some philosophical progress in thinking about species. First is the distinction between theoretical and operational concepts. If different concepts function in different ways, we can then distinguish the different criteria for evaluating concepts. We need not, for instance, require that all concepts be *operational* - used to identify and individuate species taxa, or *theoretical* - based on the theoretically important features of species taxa. And progress might also be found in the analysis of the descriptive content of species concepts. Some of this descriptive content might be associated with the definitional core, and be "essential" in some sense to our understanding of species. This definitional core, unlike the descriptive periphery, tells us what is fundamental and central to our understanding of species. What seems to be fundamental about species, even for those who disagree about species concepts, is that they are extended in time and space, and change over time. This temporal element is clear in tree thinking, where species are branches and nodes on the evolutionary tree. Some of the descriptive content, on the other hand, may be peripheral in the sense that it applies to some instances of species but not others. It might be, for instance, that reproductive isolation is relevant to some species taxa, but not others.

When we think of the definitional core we are thinking about what we believe to be most fundamental to species given our understanding of evolution: What it is that *makes* something a species, and not just what we associate with species. Analogously, when we think about the definitional core of water, we are thinking about what makes something water – its particular molecular constitution of hydrogen and oxygen, not the facts that it is necessary for life, used in swimming pools, falls from the sky as rain, and covers a majority of the Earth's surface. This descriptive periphery may be important practically, but it doesn't get at what is most fundamental about our conception of water. Similarly, the descriptive periphery of a species concept may be important in many ways, but it doesn't get at what is most fundamental about species. For that we can think about species, and biological taxa in general, in *metaphysical* or *ontological* terms.

The term 'metaphysics' has a problematic history. In contemporary popular usage it is sometimes associated with paranormal phenomena – mind reading, communication with the dead, and the exploration of spiritual realms. And among scientists it is sometimes used to refer to useless, abstruse thinking that has little to do with the actual world. If metaphysics is either of these things, then the metaphysics of biological classification is hardly worth our time and effort. The origin of the term 'metaphysics' seems to be from Andronicus's compilation of Aristotle's works, where *The Metaphysics* was simply the set of writings that came after *The Physics* – "te meta ta phusika." Aristotle's descriptions of his projects in *The Metaphysics* are usually translated as "first philosophy" or "first science." Topics addressed include "being as such" and "first causes." Why these writings appear after *The Physics* is not clear. It may be just that they were regarded as more difficult than those of *The Physics* and should therefore be studied after *The Physics*.

Contemporary metaphysics, as taught in universities, often seems to be a more or less unconnected set of topics: determinism and the possibility of free will; modal thinking, necessity, and contingency; the nature of properties and objects; the nature and reality of time; and the existence of God. There is also, though, a way of thinking about metaphysics that is more coherent and relevant to science. According to this approach, metaphysics is an exploration of the general features of the world, as Simon Blackburn explains:

> Metaphysics is the exploration of the most general features of the world. We conceive of the world about us in various highly general ways. It is orderly, and structured in space and time; it contains matter and minds, things and properties of things, necessity, events, causation, creation, change, values, facts and states of affairs. Metaphysics seeks to understand these features of the world better. It aims at a large-scale investigation of the way things hang together. (Blackburn 2002, 61)

Metaphysics, in this sense, does not go beyond the world we experience, but addresses the things in the world at a higher level of generality. Given the specific things that we find in the world, is there some more general way to think about these things? This is the sort of thing Descartes was doing when he argued that there were two kinds of substances, an extended material substance and a nonextended thinking substance.

The example of Cartesian metaphysics seems to count against a turn to metaphysics in science though. It isn't obvious that a nonextended thinking substance could even be the subject of scientific investigation. We can avoid this possibility though if we adopt a naturalistic starting point that tells us to begin with what science tells us about the world. With a naturalistic metaphysics, we start with science and then proceed to a more general account of nature. We can follow Newton's example when he proposed that we should think about gravity as a *force*. We can think about gravity specifically, and how it explains planetary motion, the descent of stones on earth, and the tides. But also, given what we know about gravity and how it works, what is it at a more general level? At a more general level it is a *force* – a basic kind of thing in nature. In treating gravity more fundamentally as a force we are engaged in a naturalistic metaphysics.

W. V. O. Quine, one of the most prominent advocates of naturalistic metaphysics in the twentieth century, argued that this sort of metaphysics is not separate from science, but rather on a continuum with science. He asked us to think about metaphysics in terms of the ontological presuppositions of our theories and the language we use to express them. According to Quine, these ontological commitments – about what fundamental kinds of things exist, are just as much a part of science as the more directly empirical claims about what we observe and what constitutes our theories.

> Within natural science there is a continuum of gradations, from the statements which report observations to those which reflect basic features say of quantum theory or the theory of relativity. The view which I end up with ... is that statements of ontology or even of mathematics and logic form a continuation of this continuum, a continuation which is perhaps yet more remote from observation than are the central principles of quantum theory or relativity. The differences here are in my view differences only in degree and not in kind. (Quine 1976, 211)

If there is a continuum, as Quine argued, then metaphysical questions are not separate, or external to science.

A naturalistic metaphysics can be either descriptive – telling us simply what basic sorts of things are implied by science, or prescriptive – telling us how we should conceive the basic sorts of things. The prescriptive approach is typically revisionary in that it is asking us to reject an old metaphysics and adopt a new. But often the two approaches are combined, in the sense that the revisionary account begins with the descriptive, as Blackburn explains:

> Metaphysics may be a purely descriptive enterprise. Or, it may be that there is reason for revision: the ways we think about things do not hang together, and some categories are more trustworthy than others. Revisionary metaphysics then seeks to change our ways of thought in directions it finds necessary. The distinction between revisionary and descriptive metaphysics is not sharp, for it is out of the descriptions that the need for revision allegedly arises. (Blackburn 2002, 61)

We will be adopting this approach here, starting with a descriptive metaphysics, how we actually think about biological taxa and categories based on evolutionary theory, but we will also explore some of the reasons for adopting or rejecting particular metaphysical stances. Since most of the recent debates about metaphysics have been about the species category, we will start there and then look at other levels of biological classification.

The Metaphysics of Species

In Chapter 6 we saw that species taxa, as we typically identify and individuate them, can vary greatly across biodiversity. Big sexually reproducing vertebrate species are very different from the species we see among invertebrates, plants, fungi, and bacteria. But these differences do not imply that there are no similarities in species across biodiversity. The insight of the hierarchical pluralism discussed in the Chapter 6 is that all of these differences among species taxa are captured by the many *operational* species concepts that help us identify and individuate species. But also on this hierarchical approach, there are universal features of species. These features are represented by *theoretical* species concepts. According to de Queiroz and Mayden, evolutionary theory implies that species taxa are spatiotemporally extended. They have beginnings and endings, with change over time. They extend across space as populations that may interact, but are still sometimes geographically isolated from other species populations. On this way of thinking species are, at a fundamental level, segments of population lineages, with beginnings in speciation and endings in speciation or extinction. This is how they have been so typically been represented on evolutionary trees. If so, how should we think about them metaphysically? What basic kinds of things are they?

In recent years, debate has focused on two main metaphysical positions – species-as-sets and species-as-individuals. On the species-as-sets approach, species consist of sets of organisms that are members of a set by virtue of their possession of one or more properties. The relation here is a membership relation. The numbers 2, 4, and 6 are *members* of the set of even numbers, based on the property of being "divisible by 2." Analogously, on this sets approach, individual humans are members of the set of *Homo sapiens*, based on the fact that each of these humans has the properties necessary and sufficient to be in that set. By contrast, on the species-as-individuals approach, species are spatiotemporally located individual things, much like organisms, with beginnings, endings, and change over time. Here, the relation is *mereological*, a part–whole relation. Organisms are *parts* of the species individual, rather than members of the set. An individual dog is a part of the species individual *Canis familiaris*, and an individual human is a part of the species individual *Homo sapiens*. As we shall see, these are two very different ways of thinking about species. And later in this chapter we will extend these two ways of thinking beyond species to higher-level taxa. But first let us look at the various species-as-sets approaches, based on the natural kinds, cluster kinds, and historical kinds. Then we will turn to the species-as-individuals approach.

Natural Kinds Framework

Perhaps the most prominent version of the species-as-sets approach is the natural kinds framework we looked at in the first chapter. This is the approach identified with Plato's metaphor in the *Phaedrus* that we should divide the world up as a good butcher divides – by cutting at the joints. In Chapter 1 we identified three different kinds of kinds: *natural kinds* that are independent of human beliefs, biases, preferences, and conventions; *conventional kinds* that are dependent on human preferences and conventions; and *artificial kinds* that are arbitrary in that they are neither independent of human preferences nor the products of actual human conventions. For each of these kinds of kinds, something is that kind by virtue of possession of some particular property or set of properties. Something is the natural kind *hydrogen* if it has the atomic number 1. Something is the conventional kind *dollar bill* if it is a piece of paper of the right size, composition, and markings. And something is the artificial kind *blue food* if it is blue and a food. In each of these cases the thing is a *member* of a particular *set* of things – hydrogen, dollar bills, or blue foods, by virtue of having a particular property or set of properties.

On one standard way of thinking about these kinds of kinds, membership in each kind is determined by a set of necessary and sufficient conditions. For instance, a particular atomic number is *necessary* to be a member of some element. For hydrogen, that atomic number is 1. Anything with a different atomic number cannot be hydrogen. Having that atomic number is also *sufficient*, so anything with that particular atomic number is hydrogen. Similarly, the necessary and sufficient conditions for being a blue food are being blue and being a food. Anything with these properties is a member of the set of blue foods. In each of these cases, and especially for the natural kinds, these sets of necessary and sufficient conditions have been thought of as "essences." They are what make each thing the kind of thing it is. They are what make it a member of the relevant "kind" set.

On this way of thinking, natural kinds are timeless, unchanging, and discrete. Because the set of necessary and sufficient conditions are timeless, applying at any time and at any place, then the sets of things established by these conditions are timeless, applying at any time and any place. Because the set of necessary conditions does not change, the sets of things established by these conditions do not change. And finally, because the necessary and sufficient conditions are discrete, the sets of things are discrete.

As we saw in Chapters 2 and 3, even the pre-Darwinian naturalists did not typically think about species as natural kinds in this sense - in spite of the misleading essentialism story fabricated in the twentieth century. And after Darwin and evolutionary theory it is even harder to think about species in this way. First it isn't clear that we can identify a set of singly necessary and jointly sufficient properties with any particular species, and for obvious reasons. If species are segments of lineages that undergo at least some change over time, then the distinctive properties of a particular species will change over time. And because species are also populations that vary geographically and grade into other species, the set of distinctive properties will also vary at a particular time and across space. But even if it were possible to identify some set of necessary and sufficient properties - an essence - of a particular species, this way of thinking still seems out of step with the fact that species are spatiotemporally located, have beginnings and endings, vary within and across populations, and change over time. The bottom line is that this natural kinds way of thinking is asking us to think about historical changing things as if they were ahistorical and unchanging.

There is, though, a tradition in philosophy of thinking about species this way. In the second half of the twentieth century Hilary Putnam (1973) and Saul Kripke (1972) argued for an essentialist approach to natural kinds based on causal semantics. According to the Kripke–Putnam approach, we point to and name something. This establishes the *extension* of the term - what it applies to in the world. Science then proceeds to determine the essential properties of that thing. For example, the natural kind term 'water' gets applied to certain things in the world. Chemists then discovered that what it gets applied to is a compound of two hydrogen atoms and one oxygen atom. It is H_2O. This is the necessary and sufficient "essence" of water. Anything that has this molecular composition is water. Anything that does not cannot be water. Kripke and Putnam also applied this essentialist approach to biological kind terms, such as 'elm' and 'tiger,' that get applied to the things in the world. The reference of these terms gets set by pointing at something in the world, and then through scientific investigation we discover the essences of these things - their necessary and sufficient properties. We find out what makes an elm an elm and a tiger a tiger.

David B. Kitts and David J. Kitts followed the lead of Putnam and Kripke in advocating an essentialist approach to biological taxa as natural kinds:

> The property which all the organisms of a species share and which ultimately accounts for the facts that they cannot be parts or members of any other is not some manifest property such as the pigmentation of a feather. It is an underlying trait. Putnam is not far from the mark concerning the essential nature of lemons when he says, "What the essential nature is not a matter of language analysis but of scientific theory construction; today we would say it was chromosome structure, in the case of lemons, and being a proton-donor in the case of acids." (Kitts and Kitts 1979, 617–618)

Kitts and Kitts concluded: "Biologists search for the underlying trait which explains the necessary relationship between an organism and its species in the genetic structure of the organism" (Kitts and Kitts 1979, 618). More recently, Michael Devitt has argued for a similar version of essentialist natural kinds thinking, claiming that "Linnaean taxa have essences that are, at least partly, intrinsic underlying properties" (Devitt 2008, 346). Like Kitts and Kitts, he thinks the essences of species are to be found primarily in the genetic properties.

> In sexual organisms the intrinsic underlying properties in question are to be found among the properties of zygotes; in asexual ones, among those of propagules and the like. For most organisms the essential intrinsic properties are probably largely, although not entirely, genetic. Sometimes these properties may not be genetic at all but in "the architecture of chromosomes," "developmental programs," or whatever ... For convenience, I shall often write as if the essential intrinsic properties were simply genetic but I emphasize that my Essentialism is not committed to this. (Devitt 2008, 347)

Devitt looks for essences here because that is where he thinks some central explanatory power lies. Essences are largely genetic, because genetics explains the other properties of organisms, and allows us to generalize about these organisms (Devitt 2008, 352). There is an initial plausibility to this way of thinking. Surely there is some set of genetic traits associated with a species that is distinctive to each species and can explain our generalizations about members of that species. Tigers have stripes, and this is explained by a set of genes and regulatory networks that cause tigers to develop stripes.

But as we have seen in the history of classification, this essentialist natural kinds approach does not reflect the actual practice of systematists when grouping organisms into species. An individual organism is usually taken to be a member of a particular species if it is part of a particular segment of that population lineage. Something is a tiger *because* it was born of tigers – whether or not it has tiger-typical stripes and coloration. And systematist are aware of the variability within populations and in a lineage over time. Sexual dimorphism in some species is an especially striking source of variability. Recall from Chapter 6 the barnacle species, *Trypetesa lampas*, where the male is much smaller and parasitic on the female, lacking even a gastrointestinal tract. Devitt acknowledges this variability, but dismisses it, telling us that "an intrinsic essence does not have to be "neat and tidy'" (Devitt 2008, 371). But the essences of natural kinds are supposed to be "neat and tidy" in an important way – as a single set of necessary and sufficient properties. This is one of the virtues of the natural kinds approach. The essence of water for instance, as H_2O, is the epitome of neat and tidy. If the essences of biological species are not neat and tidy in a similar way, they may not be natural kinds in the same way.

The difficulties of such an essentialist approach become even more obvious when we try to understand how evolutionary change might here be accommodated. Natural kinds on this approach are timeless and eternal because the set of properties that make an organism a member of a natural kind are also timeless and eternal. Change from one kind to another is possible in principle even though these kinds are eternal. We could, for instance, change a substance from water into oxygen and hydrogen. And we might even change one elemental substance into another. But evolutionary change seems different. It seems to imply a gradual change in the properties of organisms rather than a discontinuous leap from one kind to another. This raises the worry that species cannot evolve if they are natural kinds, and if they evolve they cannot be natural kinds. Devitt has a response to this worry though, arguing that even though a natural kind itself cannot evolve, because its essential properties are timeless and unchanging, a lineage of organisms can still change, by passing from one kind to another.

> Suppose that S1 and S2 are distinct species, on everyone's view of species, and that S2 evolved from S1 by natural selection. Essentialism requires

> that there be an intrinsic essence G1 for S1 and G2 for S2. G1 and G2 will be different but will have a lot in common. This picture is quite compatible with the Darwinian view that the evolution of S2 is a gradual process of natural selection operating on genetic variation among the members of S1. (Devitt 2008, 372)

This suggests that in evolutionary change from S1 to S2, with respective and distinct essences G1 and G2, there is a leap from one natural kind to another, even if the actual change within the lineage was gradual. If so, then at some time a group of organisms will be in between essences, not fully having either G1 or G2.

> On the Essentialist picture, the evolution of S2 from S1 will involve a gradual process of moving from organisms that determinately have G1 to organisms that determinately have G2 via a whole lot of organisms that do not determinately have either. There is no fact of the matter about where precisely the line should be drawn between what constitutes G1 and what constitutes G2, hence no fact of the matter about where precisely to draw the line between being a member of S1 and being a member of S2. Essences are a bit indeterminate. (Devitt 2008, 373)

Devitt's admission here is surely worrisome. It implies that there might be many organisms that are not strictly a member of a species, because they don't strictly have the essences associated with that species. What status do they have then? If this is a common situation, it is surely problematic for this version of essentialism.

This essentialism also demands an account of how the essences G1 and G2 are determined. Are they determined by observation of the organisms in a lineage? If so, at what time should we observe these organisms? In gradual change a group of organisms will vary over time. They will mostly have one set of properties at one time and another at a later time. We will have to decide which time frame is appropriate for determining the essence. But it isn't clear how we should do this. If the essences are not determined by observation, then we need an account of how some other approach will be satisfactory for an empirical science. Perhaps one could be clever enough to respond to all these problems, but the fact will still remain that there is discordance in the attempt to conceive changing things in nature by a metaphysical account that does not seem to accommodate change.

Cluster Kinds

The essentialist version of natural kinds is a conjunctive approach based on the idea that essences will at least sometimes be a conjunction of necessary properties that together are sufficient for making a thing the kind of thing it is. Whereas some essences, such as atomic number or molecular composition seem to involve only a single property, in biological taxa, there will surely be many more. Stripes are distinctive of *Panthera tigris*, for instance, but so are size, behavior, mating tendencies, and more. Presumably there will be many properties that distinguish tigers from lions and other cats. Similarly, the essence of humans, on this approach, would be a conjunction of many properties that would together distinguish members of the set *Homo sapiens* from members of the sets *Pan troglodytes* (chimps) or *Pan paniscus* (bonobos). In times past, this conjunction might have included language, reason, and tool use.

Alternatively, there may be no single *conjunctive* set of necessary and sufficient conditions to distinguish tigers from other cats and humans from other primates. Perhaps there is instead a *disjunctive* set of properties that all organisms have that make them a member of a particular species. Essences here would be a disjunction – some combination or other – of a set of properties. The essence of a kind would then be a subset or "cluster" of some larger set of properties. One individual can have one subset or cluster of properties and be a member of *Panthera tigris*, while another individual can have a different subset or cluster and also be a tiger. This is the idea of family resemblance attributed to Ludwig Wittgenstein in his *Philosophical Investigations*. Two members of a family might resemble each other in one way, while there might be different ways in which they each resemble other members of the family (Wittgenstein 1968, 31). There is, then, a cluster of traits distinctive of this family, but not all members have exactly the same subset of these traits.

This idea of a cluster kind was first applied to biological classification by Morton Beckner, but has since been adopted by Richard Boyd (1999), R. A. Wilson (1999), Matthew Slater (2013), Muhammad Ali Khalidi (2013), and more. The most prominent of these approaches is the "homeostatic property cluster" or HPC approach. There are two components to this view. First is the disjunction of clustered properties, as Robert Wilson explains:

> The basic claim of the HPC view is that natural kind terms are often defined by a cluster of properties, no one or particular n-tuple of which must be possessed by any individual to which the name applies, but some such n-tuple of which must be possess by all such individuals. (Wilson 1999, 197)

This approach allows for variability within a species from sexual dimorphism, stages of development, and more. Boyd argues we can think about this variability in terms of "conditionally specified dispositional properties."

> The fact that there is substantial sexual dimorphism in many species and the fact that there are often profound differences between the phenotypic properties of members of the same species at different stages of their life histories (for example, in insect species), together require that we characterize the homeostatic property cluster associated with a biological species as containing lots of conditionally specified dispositional properties for which canonical descriptions might be something like, "if male and in the first molt, P," or "if female and in the aquatic stage, Q." (Boyd 1999, 165)

The females of a species, for instance, would be associated with some subset of properties, while males would be associated with some other subset of properties. And being at some stage of development for each would be associated with a distinctive subset of properties.

The second component of the homeostatic property cluster approach is the set of causal mechanisms that are responsible for the clustering of properties in a species.

> A variety of homeostatic mechanisms – gene exchange between certain populations and reproductive isolation from others, effects of common selective factors, coadapted gene complexes and other limitations on heritable variation, developmental constraints, the effects of organism-caused features of evolutionary niches, and so on – act to establish the patterns of evolutionary stasis that we recognize as manifestations of biological species. (Boyd 1999, 165)

Because of gene exchange and reproductive isolation, for example, there will patterns of shared traits within a species. Similarly, the operation of natural selection in some ecological niche will produce stability in some traits within a species. What is important here is that there are a variety of processes that operate within a species and that produce a relatively stable set of properties in the members of that species, so that each member will have some subset or other of those properties.

The obvious advantage of the cluster approach of the essentialist natural kinds approach is that it can accommodate variability within a species, as long as we can identify some stable subset of properties associated with each variation. The cluster of traits associated with males and females will be different, but will be stable among males and females. There will be stable clusters of traits among those individuals in particular environments, at high elevations perhaps. And there will be stable clusters of traits among those individuals at particular stages of development.

This clustering approach seems to work well when we have *already* identified and individuated species, but it isn't obvious how we can use this approach to group organisms into species if we don't already know what group of organisms constitutes a species. The important question is this: Which clusters are relevant to being a member of a particular species? We could find clusters of properties among any group of organisms related to size, color, and so on. But only some of these clusters will be relevant to species groupings. Perhaps reference to homeostatic mechanisms will help, but not all homeostatic mechanisms operate just within a species. Natural selection, for instance, may operate similarly across many species, relative to thermoregulation, predation avoidance, and so on. Further guidance here is needed.

But also it isn't obvious how a cluster approach can accommodate evolutionary change. Natural selection is not always a stabilizing force. If it were, we would not have evolution by natural selection. How do we accommodate this change within a species through a cluster approach to sets? If, as with the conjunctive sets of properties in the essentialist natural kinds way of thinking, these disjunctive sets of properties are timeless – they apply anytime, then it isn't obvious how cluster thinking can accommodate change over time. Wouldn't the cluster of properties associated with different members of a particular species need to change as the species lineage changes? Females within a species may change of time, as will the developmental stages the females pass through. But then the cluster of properties would not be determining membership into a species set. It would simply be a description of some subset of those organisms in a particular species at that time. Ultimately the problem here is the same problem we saw in essentialist natural kinds. We are trying to represent something that changes over time with a metaphysical framework that does not obviously accommodate change. Is there some other way to think about species more generally that can accommodate the facts that species have histories and that these histories involve change? A third way of thinking about species as sets – as historical kinds, attempts to do so.

Historical Kinds

Both the essentialist and cluster thinking about species assume the properties that determine membership in a species kind set are intrinsic. *Intrinsic properties* are roughly those properties that don't require reference to anything external to the organism. Typically this includes morphological properties such as size, shape, and color, or the genetic properties that guide development. *Extrinsic properties*, by contrast, require reference to something external – a relation to something else. Being "a parent of" or "larger than" are both extrinsic properties, because they depend on something external, the presence of an offspring or some other physical object. This distinction is perhaps not fully clear and unambiguous, as some properties may be conceived both intrinsically and extrinsically. We can, for instance, treat color as an intrinsic property of objects. But we can also treat it as a relation between a subject, object, and photic environment. After all, what color a thing has is dependent on the light in the environment and the perceptual apparatus of a perceiver (Richards 2005). The colors of flowers, for instance, may be very different for those organisms that are sensitive to ultraviolet light than they are for those that are not. Color in this way of thinking is not intrinsic. Nonetheless, the distinction between intrinsic and extrinsic properties (or relations) may still be useful.

Perhaps we can then think about species sets as being determined by extrinsic properties or historical relations rather than intrinsic genetic or morphological properties. As we have seen, the main problem for essentialist and cluster approaches is that they do not obviously accommodate the historical component of species. But as we have seen in the history of classification, it has been standard to think about species historically as a lineage of ancestor–descendant organisms. Aristotle, Linnaeus, Buffon, and

Darwin's contemporaries all thought this way – even though none of them were evolutionists in a modern sense. And since Darwin, thinking about species has been explicitly historical in the construction of evolutionary trees. Generally speaking, an organism is a member of a species if it is descended from members of that species. Charles Darwin is a member of *Homo sapiens*, because his parents were members, and if he were not born of human parents, he could not be a member of *Homo sapiens*. Perhaps this historical ancestor-descendant relation is essential to species inclusion.

This view is suggested by Michael Ruse, when he asked why his dog Spencer is a member of the species *Canis familiaris*: "So why do we want to say that he is part of the species? Because he descended from the original ancestors, along with the rest of the group – that's why ... Descent is starting to look very much like an essential property" (Ruse 1987, 236). Paul Griffiths also argues for a historical essentialism:

> Nothing that does not share the historical origin of the kind can be a member of the kind. Although Lilith might not have been a domestic cat, as a domestic cat she is necessarily a member of the genealogical nexus between the speciation event in which the taxon originated and the speciation or extinction event in which it will cease to exist. It is not possible to be a domestic cat without being in that genealogical nexus. (Griffith 1999, 219)

On this approach, the essence of a species is its location within the evolutionary tree. If an organism has the relevant relation to other organisms on the tree of life, and this relation helps to explain the important features associated with a species, then it is a member of that species set (See also LaPorte 2004, 64.)

The advantage of this historical essentialism as the basis for species sets is that it reflects the fact that species are historical. They are lineages of ancestors and descendants. It can also straightforwardly accommodate change. The crucial properties are not those intrinsic morphological or genetic properties that change over time. They are the relations an organism has with other organisms, a relation that does not change as the intrinsic properties change. In this way, historical kinds better reflect actual biological practice. But there is a problem with this way of thinking as well. Species are not the population lineages themselves, but *segments* of lineages. Species have beginnings and endings. Consequently it is not quite accurate to say that the historical ancestor–descendent relation is always determinative. There are breaks in the lineages between different species. At some point in time, the members of a lineage cease to be members of one species, and come to be members of another. How do we identify those breaks that segment a lineage and indicate a new species has formed? A simple historical essentialism that determines species set inclusion only on the basis of ancestry will not do so. Ancestor-descendant relations continue over speciation events just as they do within a species lineage. In other words, historical relations may be necessary for species membership, but are not sufficient. Some historical relations will determine species set inclusion, but some will not. Something additional is needed for the historical kinds approach to reflect actual practice in biological classification. That additional something might plausibly require reference to the processes that result in speciation. But how to do this on a historical kinds approach is unclear.

Perhaps we should instead think about species as sets, but not under any single set conception. Perhaps we should be *pluralists* about species sets. This sort of pluralism is implicit in a recent approach to natural kinds thinking. According to Muhammad Ali Khalidi natural kinds are really just *epistemic kinds*, "categories that enable us to gain knowledge of reality" (Khalidi 2013, xi). Epistemic kinds are whatever categories each of the sciences uses to successfully explain and predict phenomena. He does not assume that all natural kinds, from physics and chemistry to biology and the social sciences, have the same features. There might be, in other words, multiple kinds of natural kinds. P. D. Magnus argues similarly for a pragmatic account of natural kinds, whereby "a category of things or phenomena is a *natural kind* for a domain if it is indispensable for successful science of that domain." And what is indispensable may vary across domains (Magnus 2012, 2).

This sort of pragmatic pluralism might also be adopted *within* a single domain. Philip Kitcher for instance:

> Species are sets of organisms related to one another by complicated, biological interesting relations. There are many such relations which could be used to delimit species taxa. However, there is no unique relation which is privileged in that the species taxa it generates will answer to the needs of all biologists and will be applicable to all groups of organisms. (Kitcher 1992, 317)

According to Kitcher's approach, biologists can legitimately divide a single group of organisms into different sets based on a variety of "structural" features based on genetic, chromosomal, or developmental traits, or on "historical" criteria. Perhaps some of these sets may be determined by necessary and sufficient properties, some on the basis of a cluster of properties, and others historically on the basis of genealogical relations. But as we saw in Chapter 6, these sets of organisms do not necessarily correspond to what we would identify as species. Moreover, adopting a pluralism about sets does not eliminate the seeming discordance between timeless sets thinking and the historical, changing nature of species. If *none* of the sets conceptions seem to adequately represent the historical nature of species, then a sets pluralism just allows for a choice of inadequate approaches.

Species as Individuals

The problems with these sets approaches – their apparent difficulties representing variation, evolutionary change, and speciation, do not by themselves refute the sets metaphysics. A sets approach might still be the best way of thinking about species. There is an alternative though, one that has been finding increasing support in both biology and philosophy. Perhaps we should think about species as *individuals* with organisms as parts, rather than as *sets* with organisms as members. Michael Ghiselin proposed this idea while serving a postdoctoral fellowship under Ernst Mayr in the 1960s, and then developed it in his 1969 book *The Triumph of the Darwinian Method*. He argued for a "moderate nominalism" based on the idea that species are potentially interbreeding populations that are reproductively isolated from other groups of organisms. This suggests, according to Ghiselin, that species taxa have a sort of cohesion lacking in other groups of organisms, and that species constitute an integrated level of organization above the level of the organism (Ghiselin 1969, 53–54).

This higher level of organization is perhaps most striking when we look at our own species – *Homo sapiens*. It is not just that humans share certain features that might group them into a set – bipedalism, opposable thumbs, language, and reason; they also recognize each other as the same sort of creature, look upon each other as potential mates, and sometimes interbreed. Humans also exchange information, gather in communal activities, dance and sing together, and do all this largely to the *exclusion of members of all other species*. We humans may be similar in a variety of ways, but we also share our lives with other humans in ways that we don't share with other, nonhuman, animals. There is, then, *cohesion* among humans based on these social processes. And we see a similar sort of cohesion among other species. Members of a bird species reproduce sexually, identify each other visually and through song, cooperate to raise young, and fly together in flocks. This isn't to say that members of one species never interact with members of another species though – coevolution has produced many examples, but this cohesion through social interaction is surely stronger among conspecifics than it is among organisms of different species.

In this 1969 book, Ghiselin also argued more generally that there are just two basic, fundamental kinds of things: individuals and classes (or 'sets' in the terminology here). Individuality is found in many sorts of things, and at all levels, biological and otherwise:

> In the usual biological sense, 'individual' is a synonym for 'organism' but the ontological term is a much, much broader one. Although all organisms are individuals in the ontological sense, not all individuals in the ontological sense are individuals in the usual biological sense. We have suggested all sorts of things lacking the defining properties of 'organism' that might be given as examples of an ontological individual. A chair is a piece of matter that an organism might sit on, and the world is full of such things. Or consider a part of an individual: one of a person's legs, or one of the legs of a chair. A part of an organism can be individual, including not just each and every organ, but each and every cell, each and every molecule, and each and every atom ... Likewise we can say that larger things can be individuals. An individual society would be a good example. If you do not like a society as an example of an individual, try the Earth, the Solar System, the Milky Way, and the Universe. (Ghiselin 1997, 37–38)

On this view, individuality extends above and below the level of individual organism, from cells to the solar system and the universe itself.

We can understand this species-as-individuals idea by analogy with organisms. First, an organism is spatiotemporally located, with a beginning and an ending and continuity in between. Similarly, a *species* individual is located in time and space. It has a beginning and an end, and a specific location in space.

> [A]n individual occupies a definite position in space and time. It has a beginning and an end. Once it ceases to exist it is gone forever. In a biological context this means that an organism never comes back into existence once it is dead, and a species never comes back into existence once it has become extinct. And although it might move from one place to another, there has to be a continuity across space as well as through time. (Ghiselin 1997, 41)

Second, just as individual organisms have parts and not members, a species has parts and not members. The head or heart of Socrates, for instance, is not a member of Socrates, but is instead a part. Organisms are analogously *parts* of species (Ghiselin 1997, 38). Third, individuals are concrete, whereas sets (or classes) are abstract. Individuals can participate in processes and play causal roles. An individual human is a concrete thing and can do things, or have things done to it. Analogously a species individual can do things and have things done to it (Ghiselin 1997, 43). But a *set* of organisms cannot do anything or have things done to it, and cannot participate in processes. Finally, individuals are not the subject of scientific laws.

> [T]here are no laws for individuals as such, only for classes of individuals. Laws of nature are spatio-temporally unrestricted, and refer only to classes of individuals. Thus, although there are laws about celestial bodies in general, there is no law of nature for Mars or the Milky Way. Of course laws of nature apply to such individuals; they are true, and true of physical necessity, of every individuals to which they apply. (Ghiselin 1997, 45)

So there are no laws of *Homo sapiens*, or any other species taxon, any more than there could be a law of Socrates or of a particular electron or planet. (This is not to deny, of course, that the physical laws of nature apply to the individual organisms that make up a species function.)

This species-as-individuals thesis has come to be accepted by a significant number of biologists and philosophers of biology. (See Mayden 1997, de Queiroz 1999, Brogaard 2004, Richards 2010. For one recent criticism of this thesis see Slater 2013.) One of the most prominent advocates was the philosopher David Hull, who was initially critical, but came to support the individuals thesis, in part because it made sense of the fact there seemed to be no laws of biological taxa such as *Homo sapiens*. Like Ghiselin, Hull treated individuals as spatiotemporally localized, cohesive, and continuous entities. And also like Ghiselin, he extended this idea beyond species and biology. According to Hull, social groups, theories, and concepts should also be conceived as individuals (Hull 1992, 294).

Individuality

The obvious advantage of the species-as-individuals approach is that it seems to reflect the historical nature of species taxa. Evolutionary theory tells us that like organisms, species are spatiotemporally located things, with beginnings, endings, continuity, and change over time. If we adopt a naturalistic approach to metaphysics that begins with what science tells us about the world, then the species-as-individuals approach has at least something to recommend it. It seems to follow from the scientific way of thinking about species as historical, changing things. But the problems of the species-as-individuals approach are apparent as well.

The most obvious problem is the disanalogy with organisms. Species seem to lack the integrated causal cohesion we see in organisms. In vertebrates, for instance, lungs (or gills) and hearts interact as part of a cardiovascular system. Muscles are integrated with skeletal structure. Gastrointestinal systems extract chemical nutrients from food, provide these nutrients to all the other systems, and eliminate waste products. Vertebrates require this causal integration to live. Disrupt any of these systems and death is likely. Species, by contrast, seem to have much less causal integration and can survive extensive disruption. There are no causally integrated physiological systems, and the individuals of a species can be separated or destroyed, without the destruction of the species itself. Michael Ruse found this disanalogy problematic:

> We think organisms are individuals because the parts are all joined together. Charles Darwin's head was joined to Charles Darwin's trunk. But in the case of species, this is not so. Charles Darwin was never linked up to Thomas Henry Huxley. Of course, you might object that although Darwin's head was never linked directly to his feet, they were linked indirectly through intermediate parts. Analogously, as evolutionists presumably we believe that Darwin and Huxley were linked by actual physical entities (namely the succession of humans back to their shared ancestors). But, this objection fails, for the point is that these links have now been broken and lost. If (gruesome thought!) Darwin's head were physically severed from his feet, we would certainly have no biological individual. (Ruse 1987, 232)

Surely we should take this criticism seriously. Even the most integrated species seem far less integrated than organisms. If so, this is a big problem for the species-as-individuals thesis. After all, physiology is part of science as well, and if it seems to militate against the species-as-individuals thesis, then a naturalistic ontology suggests species cannot be individuals in a manner similar to organisms.

Philosophers typically find this cohesion objection more compelling than do many biologists. In part this may be a consequence of a difference in perspective. When philosophers think of organisms, they often think about the most familiar of organisms – large vertebrates such as humans, dogs, and horses. And typically what they are most interested in are humans. Human identity and individuality have long been the subject of philosophical debate. Philosophical analysis of individuality has therefore typically started with human or similar organisms. Biologists, on the other hand, typically consider the full range of organisms from humans and big mammals to birds, insects, fungi, and bacteria. David Hull noted this difference in the philosophical and biological perspectives, and endorsed the biological:

> Differences between these two analyses have three sources: first, philosophers have been most interested in individuating persons, the hardest case of all, while biologists have been content to individuate organisms; second, when philosophers have discussed the individuation of organisms, they have usually limited themselves to adult mammals, while biologists have attempted to develop a notion of organism adequate to handle the wide variety of organisms which exist in nature; and finally, philosophers have felt free to resort to hypothetical science fiction examples to test their conceptions, while biologists rely on actual cases. In each instance, I prefer the biologists' strategy. A clear notion of an individual organism seems an absolute prerequisite for any adequate notion of a person, and this notion should be applicable to all organisms, not just a minuscule fraction. (Hull 1992, 301)

This difference in perspective is not surprising:

> Given our relative size, period of duration, and perceptual acuity, organisms appear to be historical entities, species appear to be classes of some sort, and genes cannot be seen at all. However, after acquainting oneself with the various entities which biologists count as organisms and

> the roles which organisms and species play in the evolutionary process, one realizes exactly how problematic our commonsense notions are. (Hull 1992, 295)

If Hull is right, a scientific starting point – a naturalistic metaphysics, requires that we began by looking at the full range of biodiversity.

If we follow Hull's suggestion here, the cohesion objection seems to lose at least some of its force. Not all organisms are as cohesive as the big vertebrates we see around us, as Ghiselin explains:

> A situation in which an organism breaks up into component parts that never get back together again is familiar even to the lay person. Propagation by cuttings, budding, and fission of some animals such as starfish are good examples. There are fewer examples of organisms that break up, then fuse back together, but slime-molds are an example. These fungi, which form lineages produced by asexual reproduction, forage independently on organic materials. Later in their life cycle they come together to form a single mass, complete with reproductive organs that give rise to spores. Sometimes they are called "social amoebae," and the term aptly compares them to societies that are united only from time to time. (Ghiselin 1997, 55)

Slime molds seem to have functional integrated cohesion at some point in their life cycle, as they form a single mass with specialized "organs." But then they can break apart and function independently. This suggests that we can think of slime molds as exhibiting individuality at two levels – a higher level where the parts form reproductive organs and function as an individual organism, and a lower level where the parts separate and forage independently. If we look throughout biodiversity, beyond the more highly integrated vertebrates, we see a wide range of integration and cohesion, from what seem to be social organisms, such as slime molds and fungi, to more integrated organisms that still have social elements, such as jellyfish, to organisms that can bud or divide, all the way up to highly integrated vertebrates. If we consider this full range of integration, we might be less likely to treat the more cohesive and integrated vertebrates as paradigmatic. If so, then the relative lack of cohesion within a species may seem less problematic.

This disanalogy of species individuality with organismic individuality is therefore not as powerful as it might first seem. Individuality comes in degrees among organisms. Some organisms are more cohesive, others less so. This suggests that there is vagueness in how we think about organisms. The more cohesive will be *more* "organismic," the less cohesive less so. There also seems to be multiple criteria for determining whether an entity is an organism. For instance, we could identify organisms on the basis of functional integration or spatial cohesion, as suggested in the aforementioned slime mold example, or on the basis of homogeneous cellular genotype, where an individual organism is just that group of cells that have the same genotype (Clarke and Okasha 2013). Alternatively, we might also use immune systems to determine what counts as an organism. The basis for organismic individuality, according to Thomas Pradeu, is that the immune system monitors and responds to aberrant molecular patterns. How it responds to these patterns *determines* the boundaries of the individual organism (Pradeu 2013). Recent research into the microbiome complicates matters further. Organismic individuality might plausibly include all the bacteria that colonize a body, as well as the monogenomic cells of that body. If so, then a human individual, described as a "symbiotic superindividual" by Frédéric Bouchard, is actually composed of multiple species of organisms (Bouchard 2013). What this all suggests is that individuality is complex and problematic at the organism level, as well as at the species level.

There is another reason the cohesion objection may be less compelling to biologists than to philosophers. Biologists are typically more cognizant of the variety and complexity of cohesion processes within species. Cohesion processes can function negatively, isolating the organisms of one species from the organisms of another. Or they can function positively, in terms of processes that cause cohesion among organisms of a species. Or do both. Interbreeding and gene flow within a species, for instance, may isolate one group of organisms from another in that they don't occur across species, but they are also cohesive in that genes are exchanged among the organisms within a species. This cohesive capacity is more complex than generally recognized, however. We might assume that there is a clear distinction between sexual and asexual organisms, and that only the sexual have this capacity. But some organisms reproduce sexually under some circumstances and asexually under others. Some green algae, for instance, reproduces sexually in dry phases and asexually when hydrated (Niklas 1997, 42). If so, then cohesion is episodic, occurring at some times, but not others. And gene exchange may also occur over large distances, in the plant kingdom in particular (Niklas 1997, 27).

Cohesion may also be a product of behavior and behavioral tendencies. The mate recognition systems of individual organisms, for instance, focus on those individuals that are potential mates, and distinguish them from those that are not, based on chemical communication and behavior. Territorial and courtship displays among birds, for instance, are species specific, as are calls among primate species. The head bobbing movements of spiny lizards are species specific and allow females to identify the males of their own species and avoid the males of other species (Wilson 2000, 183–184). But not all of these cohesive factors are directly related to reproduction. Conspecifics recognize, communicate, and interact with each other in many ways that they don't with organisms of other species. This results in social spacing effects, social symbiosis, dominance, and caste systems. There are density-dependent effects, where the density of conspecifics influences emigration rates and development. At higher densities, the caterpillars of the cotton leaf worm *Spodoptera littoralis*, for instance, become darker, more active, and produce smaller adults. In some aphid species, higher density produces a shift from parthenogenesis to sexual reproduction and dispersal (Wilson 2000, 83). Density effects are perhaps most striking in plague locusts:

> When these insects are crowded during periods of peak population growth, they undergo a phase change that takes three generations, from the *solitaria* phase that is first crowded through the intermediate *transiens* in the second generation to the *gregaria* phase in the third generation. The final adult products are darker in color, more slender, have longer wings, possess more body fat and less water, and are more active. In short, they are superior flying machines. Also, their chromosomes develop more chiasmata during meiosis, resulting in a higher recombination rate and, presumably, greater genetic adaptability. Finally, both the nymphs and the adults are strongly gregarious, readily banding together until they create the immense plague swarms. (Wilson 2000, 83)

Cohesive effects can be found not just in social insects, but also in fish, frogs, birds, prairie dogs, dolphins, elephants, lions, and primates. In each of these cases, cohesion results from the fact that organisms of a species respond differentially to members of their own species.

This social cohesion is most apparent to us in our own species. Not only do we identify other humans as potential mates, but we also identify them and interact with them on the basis of language. Even if we speak different languages, we recognize that other humans have language and we can potentially communicate with them by language. We recognize that other humans have culture in the same way we do, and have cultural exchange with them, but not in general with members of other species. And even if we engage in activities with other species, dogs and cats in particular, we don't do so in the same way we sing, dance, play games, and eat with other humans. The point is this: sexual reproduction is not the only source of cohesion within a species. Social interaction, especially in highly social species, can serve to integrate and make a species cohesive.

Selection processes can also produce cohesion. In sexual selection, for instance, the mating tendencies and preferences of females affect the mating behaviors and physiology of males. Similarly, when males battle among themselves for mating opportunities, there are effects on male physiology and behavior. Natural selection in a common environment can sometimes have a cohesive effect, maintaining homogeneity and stasis within a population. Or, as we all know, it could cause change over time. Coevolution can similarly affect the cohesion of a species. Members of one species may form coevolutionary relationships with species of parasites or bacteria (if there are bacterial species). Once this happens then the fate of each species becomes intertwined with the fate of the other.

These processes are all cohesive relative to species in the sense that they differentially affect the organisms of a single species. Whether gene exchange occurs, what gets identified by a mate recognition system, or how sexual or natural selection works, depends on species identification. But, it might be objected, many of these forces of cohesion do not operate just at the species level. Organisms come grouped into breeding pairs, families, colonies, demes, populations, and metapopulations. Perhaps there are cohesive forces at each level, and cohesion is not unique to the species level. This objection is surely correct and it reinforces the claim of Ghiselin and Hull that individuality occurs at many levels of biological organization. If so, then we might wonder: Is there something distinctive about the cohesion and individuality of species? A full answer to this question is beyond the scope of the analysis here, but we can see where we might look for an answer. Perhaps the identification of species level cohesive processes and effects is a significant part of what has driven the proliferation of operational species concepts. There are ecological, interbreeding, and mate recognition concepts precisely because these are processes that bind the members of a species together. And when these processes fail to produce species level cohesion we get new species. When organisms fail to recognize each other as potential mates, reproductive isolation occurs and that allows for divergent change.

So do these cohesion processes in species successfully respond to the objection that species don't have the kind of cohesion we see in organisms? Clearly some organisms have much more cohesion and causal integration than species. Perhaps the best way to think about it is that cohesion comes in degrees. Hull makes this claim:

> Spatiotemporally organized entities can be arrayed along a continuum from the most highly organized to the most diffuse. Organisms tend to cluster near the well-organized end of the continuum while species tend to cluster near the less organized end, but as Mayr notes ... there are entities commonly classed as organisms that are no better organized than are many species. (Hull 1989, 114)

Organisms *tend* to be more cohesive, while species tend to be less so. If so, individuality comes at many levels and in many degrees.

Sets or Individuals?

The metaphysical or ontological question we asked at the beginning of this chapter was about the fundamental nature of species. What basic kinds of things does science say they are, and should we conceive of them as sets of things with members, or as individuals with parts? One problem is that at some level we can think and talk about species as either sets or individuals. Joseph LaPorte, for instance, argues that the species-as-sets and species-as-individuals approaches are compatible:

> Let it be *granted* to s-a-i theorists that the organisms of a species constitute an individual, either because they display cohesion after all or because individuality, or individuality of the relevant stripe, does not require cohesion ... Even if it is granted in this way that there is an individual whose parts are organisms of a species, it is nevertheless the case that there is a kind here as well. (LaPorte 2004, 17)

LaPorte advocates the historical kinds approach, where species are sets of things, and inclusion in the set is determined by the possession of an extrinsic property - a particular relation to other organisms in a historical lineage. LaPorte would undoubtedly be correct if he were just claiming that it is possible to think and talk about species as sets or as individual things. But that fact is of questionable value. We can, for instance, talk about love as "a red, red rose," and jealousy as a "green-eyed monster," but that doesn't make love a flower, or jealousy a monster. We can even talk about individual organisms as sets of cells. But what we really want to know is this: Given what science tells us about species, what more basic way should we think about them?

The analysis so far has indicated that the species-as-individuals approach seems more consistent with the assumption, based on evolutionary theory, that species are spatiotemporally located and restricted, and exhibit change over time. But we also know that species exhibit patterns of similarity. And while this similarity is consistent with the species-as-individuals thesis, it does seem to be better represented by the species-as-sets approach - as long as we just think about species synchronically at a single time, rather than diachronically over time. The cohesion objection, that species lack the cohesion of organisms, seems to count somewhat against the species-as-individuals thesis, although the gradations in the cohesion and individuality of organisms weaken that objection. So where does a naturalistic approach to metaphysics, an approach that begins with what science tells us about the world, tell us about ontology - the basic, most general features of biodiversity? While I side with the species-as-individuals approach (Richards 2010), based on the general discordance between an ahistorical metaphysics and the historical, evolutionary account of biological taxa represented by the tree of life, the reasons just sketched out here may not seem conclusive.

There are, however, two other considerations that we might take into account. The first is the epistemic role species play, a role typically emphasized by philosophers rather than biologists. The idea is that species *sets* help us generalize and make inferences about the world. If organisms are placed into species on the basis of the possession of some set of properties, then we can, it is presumed, generalize about all members of that species, based on the patterns of similarities. Members of a species typically have some particular suite of traits, and those traits tend to give good reasons to expect they will have other particular traits. We can infer, for instance, that a virus that causes a disease in some humans will likely cause a disease in many or all humans. The sets approach to species is certainly in the spirit of this epistemic role of species. The species-as-individuals thinking does not rule out these sorts of inferences, but it does not explicitly license them either.

There is a second consideration that may also be worth considering. Some ways of thinking are more fertile than others. They generate new hypotheses and suggest new experiments and areas of investigation. The mechanical worldview, which asked us to think about physical processes in terms of extended substance, motion and contact forces, for instance, was useful in thinking about the functioning of the heart and the circulation of blood. And Darwin asked us to think about natural selection in light of what we know about artificial selection and breeding. In both of these cases, a particular way of thinking provided heuristics for further investigation. Perhaps thinking about species as individuals, with beginnings, endings, continuity, and change over time, and cohesion processes has a similar heuristic value. It asks us to think about speciation and divergence, and change within a lineage. It asks us to think about the processes that cause cohesion within a species – gene exchange, reproductive isolation, social interaction, and the forces of natural and sexual selection. It directs us to investigate what Matt Haber and Andrew Hamilton call "cohesion generating relations" (Haber and Hamilton 2005; Haber 2013). If this is of significant value in the biological investigation of the world, then one of the values of thinking about species-as-individuals is forward looking – how it leads us to think about species, the members or parts of species and biological processes.

Higher Taxa

This chapter has so far been about the metaphysics of *species*. We started there because much of the debate about the metaphysics of classification has focused on the species level. But as we saw in Chapter 6, the species level of classification is also often assumed to be unique. Because species are the units of evolution and basal units of classification (it is sometimes assumed), they are different in important ways from genera, families, classes, and other higher-level taxa. This difference is partly reflected in the fact that evolutionary trees are generally species trees. But if species are in fact different from other biological taxa, what can our analysis of the metaphysics of species tell us about other higher taxa? Can we apply the sets or individuals approach to genera, families, orders, and classes? If so, are there reasons to prefer one approach to the other?

We can begin by asking if the sets approach will work for these higher-level taxa. But before we do so, we need to remember the problems outlined in Chapter 5 with ranking and tree thinking. As we saw in that chapter, the Linnaean ranking system is radically inadequate if ranks represent branches on the evolutionary tree. The roughly twenty or so ranks cannot possibly represent the many thousands (or more) branches on the evolutionary tree. Moreover, significant portions of the tree have reticulation – the merging of branches through introgression and hybridization, as well as horizontal connections between branches based on gene transfer. But in spite of all these problems with tree thinking, it still represents something important about evolution and the productions of higher taxa. And *if* we adopt a phylogenetic approach, the higher-level taxon *Aves* really is something like a branch on something like an evolutionary tree. And if so, it is a more inclusive branch, with sub-branches and twigs representing lower level taxa and species. So how should we think about these larger, more complex and inclusive branches?

We can, of course try to apply the sets approaches to these higher-level taxa. On the essentialist approach to natural kinds, membership in higher taxa such as *Aves* would be determined on the basis of a set of necessary and sufficient conditions. So is there a set of properties that are singly necessary and jointly sufficient to make a thing a member of this set – a bird? Why cannot we just say that having feathers is necessary and sufficient, much like having a particular atomic number is necessary and sufficient to be a particular element? If so, everything that has feathers is a bird, and everything that lacks feathers is not. This would be a timeless and unchanging essence. But it isn't clear that something like feathers can play this role as essence. The earliest birds might have lacked feathers, and an evolving branch might lose its feathers – but still be birds. Similarly for genetic essences, perhaps the earliest birds lacked a distinctive genetic sequence, or a newly evolving branch might lose that sequence. The problem here, as with species, is that an atemporal, timeless framework is being used to represent something temporal and changing – a branch on the evolutionary tree. Just as species change over time, so do higher-level taxa, as species lineages change or new species form within each taxon. Each change might require a change in the set of essential properties, whether morphological or genetic. The bottom line is this: If we adopt a phylogenetic approach, what matters for classification into higher level taxa is genealogy and ancestry – not some set of intrinsic traits. Whether or not it would be possible to identify some set of properties all birds have and non-birds lack, an essentialist natural kind seems to be the wrong sort of thing, given what evolution tells us about higher-level taxa.

Does a cluster approach fare any better? As at the species level, this approach seems to better accommodate variability than essentialism. Bird species vary quite dramatically, from hummingbirds to penguins and ostriches. Perhaps we could formulate a disjunctive set of properties that would apply to all bird species. Feathers may not be strictly necessary to be a bird, but having feathers (along with some other relevant properties) is one way to be a bird. This approach might at least allow for the variability we see in penguin and hummingbird species, but change is a problem for cluster sets here, just as it was at the species level. As old species change and new species form, then the disjunctive subsets likely need to be changed to accommodate the new properties. New distinctive traits formed through evolution may need to be added to the larger set of traits, from which the disjunctive subsets are drawn. As the branches on the evolutionary tree grow, forming new sub-branches and producing new distinctive traits, surely the cluster of relevant traits for the higher-level taxa will need to change as well. As with the essentialist approach, the cluster approach seems ill suited for a phylogenetic classification.

With historical kinds, we also see a situation similar to species. Organisms are members of a higher-level taxon, just as they are members of a species, on the basis of extrinsic properties – historical relations. An organism is a bird because it was born of birds. Here, as at the species level, the historical kinds approach reflects the fact that classification is phylogenetic and based on ancestry. But it also does not give us guidance for determining which historical relations are relevant to determining whether an organism is a bird or not. It doesn't tell us how to distinguish those organisms that are part of the relevant lineage *segment* or branch of the tree. For that we would need to refer to the processes that operate in forming the branches of the tree. Treating higher taxa as historical kinds, as with species, gives us necessary but not sufficient conditions. All members of a taxon will be connected by ancestor–descendant relations, but that fact alone does not tell us how to segment the lineages and branches into discrete higher-level taxa.

Surely it is possible to think about higher-level taxa as sets of things – as long as we don't try to be too precise. And in fact, given how we learn general terms, it might be *most* natural to think about biological taxa in general as sets of similar things (more on this in Chapter 8). But here, as with species, there is discordance between the timeless sets approach and the historical, phylogenetic ways of thinking about taxa. What about the individuals approach? Can we think about higher-level taxa as individuals the way we might think of species as individuals? After all, higher-level taxa, like species, are located in time and space, with continuity and change over time. But unlike species, higher-level taxa lack the cohesion conferred by sexual reproduction and social interaction. There may be some cohesion due to the operation of natural selection, but it will lessen as the species of the higher taxon become more diversified and divergent in form, and occupy increasingly diverse niches. Natural selection works very differently in hummingbirds and penguins, so there would be little or no selection-based cohesion between these two groups. Higher-level taxa have less cohesion than species, so the analogy with organisms seems to be much weaker.

Michael Ghiselin recognizes the differences here between species and higher taxa, but nonetheless maintains that lineages should be conceived as individuals. He distinguishes between *cohesive individuals* such as species and the merely *historical individuals* of the higher taxa that lack cohesion. Higher-level taxa – the branches on the evolutionary tree – are spatiotemporally restricted, have continuity, and are concrete rather than abstract. This makes them individuals even in the absence of cohesion (Ghiselin 1997, 55). If so, then we can apply the idea of individuality at higher taxonomic levels. These might be a different kind of individual than what we see at the species level, but it is individuality nonetheless. Perhaps cohesiveness is not necessary to be an individual. Perhaps the phylogenetic relations among species are sufficient.

But what about introgression, hybridization, and horizontal gene transfer? In each of these cases there are real connections between lineages and across branches of the tree of life. Should we then think about these connected lineages as historical individuals? If so, individuality goes beyond higher-level taxa to the connected branches on the tree, and ultimately to the tree itself. We might even treat as individuals the plant lineages that hybridize, the strains of bacteria that share genes through horizontal transfer, and any organisms that have been connected by horizontal gene transfer. The number of individuals would then be multiplied enormously. And some of these things do not seem to be individuals even in the sense that species are individuals. Does this count against the general thesis that we should conceive of all biological taxa as individuals? Some may regard this as a *reductio ad absurdum*. Any view with these absurd consequences cannot possibly be true. But we might also think that this is just a reflection of the immense complexity of organic nature. There really are these many levels of interacting organization. The advantage of the individuals approach is that it recognizes degrees of organization – individuality, and the variety of ways that things can be organized. Such an approach might inspire us to investigate further what makes something an individual. We might, for instance, investigate the full range and nature of lineages, as Matt Haber does in his attempt to understand how lineages can be found at multiple levels in a biological hierarchy (Haber 2012).

Conclusion

There are two related metaphysical projects here: a descriptive project based on what science tells us about biological taxa and the world, and a prescriptive project based on how we *should* think about biological taxa. As with the species problem, there is so far no overall consensus, although there *may* be a superficial and historical association of the sets approach with philosophers, and the individuals approach with biologists and philosophers of biology. *If* there is an association, it would not be surprising. Many philosophers cut their philosophical teeth on the essentialist natural kinds approach with its necessary and sufficient conditions. The default starting point for many philosophers (such as Kripke, Putnam, and Devitt) is to start with a natural kinds framework. For those philosophers who are committed to this framework there is an incentive to extend it as far as possible – to make it work. And if it doesn't work, the obvious response is to adjust that framework to best solve the problems. So if the essentialist approach to natural kinds doesn't work, let's try a cluster set approach. And if that doesn't work, how about a historical set approach?

Biologists in general seem less interested in the metaphysical question altogether. That also is unsurprising, as one need not explicitly commit to a particular metaphysical view to do much of biology. Systematists are more likely to think about the basic nature of biological taxa, and may be more likely to adopt an individuality approach over a sets approach. Perhaps this is due to their familiarity with the wide range and variety of cohesion processes across biodiversity. Perhaps the individuality thesis is not so implausible from a starting point that includes familiarity with less cohesive organisms such as slime molds and jellyfish, and the many cohesion forces found in nature. But also the essentialist natural kinds framework is not part of standard biological training. In fact, as may be recalled from the discussion of the essentialism story in Chapter 2, it has been anathema since Ernst Mayr set up essentialism as the bogey-man to his modern synthesis.

Suppose that the individuality approach better reflects what science tells us about the world, and is therefore a better answer to the descriptive question. But this still leaves the prescriptive question: How *should* we think about biological taxa at a more basic general level, given what science tells us about the world? One answer here is pragmatic. It depends on purposes and goals. For the philosopher whose goal is to save and extend the sets approach, then there is good reason to continue to apply and modify a sets approach. But for biologists and systematists, the goals may be different, perhaps in applying the evolutionary framework to thinking about biodiversity. If so, the individuality approaches will likely best serve these interests. That said, perhaps there are other biological interests that might also benefit from a sets approach. We can also think about organisms in terms of trophic categories such as omnivores, carnivores, and herbivores; ecological categories such as predator and prey; or physiological categories such as parasites and hosts.

One natural response to this deconstruction of the debate about metaphysics is that surely it is not *just* about individual goals and interests. If so, then the metaphysics of classification is a purely subjective activity. Each person can then do metaphysics just to satisfy his or her own idiosyncratic goals. The philosophers then need not care about how the biologists think, and vice versa. Furthermore, there is no reason one philosopher should care about what other philosophers think, or one biologist should care about what other biologists think. In Chapter 8 we will look more closely at the different ways the various metaphysical stances can function. We will be most interested in the possibility that there are objective reasons based on theory or observation to prefer one metaphysical approach to the other. If there are no good, objective reasons, then perhaps our initial worries about metaphysics return. Perhaps metaphysics really is an abstruse activity unconnected to the real world.

8 Theory and the World

Two Methodological Stances

In Chapter 7 we took a naturalistic approach to the metaphysics (ontology) of classification, starting with what science tells us about species and higher-level taxa. In effect, this is starting with our best scientific *theories*. So we began by asking what evolutionary theory tells us about species and higher-level taxa. In doing so, we were adopting a *theoretical* stance. This is the stance adopted by Darwin, the evolutionary taxonomists, Hennig, and the phylogenetic cladists. They began with evolutionary theory – as they understood it – and classified on that basis. What this meant for them is that classification would be based on phylogeny, and understood in terms of evolutionary trees. This implies that grouping should be based on those shared traits that indicate phylogeny – homologies – and not on those that don't. It also implies that biological taxa are spatiotemporally located. They have beginnings and endings, with change over time, and locations as populations in space. These theoretical assumptions about biological taxa suggest that they *might* plausibly be regarded as individuals, rather than as timeless and unchanging sets of things.

One advantage of this theoretical approach is that it is analogous to the natural kinds thinking in chemistry and physics. When we say that water is a natural kind and its essence is its molecular composition – it is by definition H_2O, we are appealing to our best chemical and physical theories. And when we identify gold as a natural kind and define it as the substance with atomic number 79, we are again relying on our best scientific theories. We are not just relying on observation. We cannot, after all, just look at water and see that it is H_2O, as we might look at it and see that it is wet. Nor can we look at gold and just see that it has the atomic number 79, as we might see that it is gold in color. The standard way of thinking about these natural kind categories and terms is that they are embedded in scientific theories. What we have seen in previous chapters is that we can start with this *theoretical stance* about biological taxa as well.

But perhaps we could instead have begun with "pure" observation, in the sense that we just look at the world and classify on that basis alone. This seems to be in the spirit of what Buffon, Adanson, de Jussieu, and de Candolle were advocating when they claimed that classification should be based on *all* similarities and differences. The pheneticists and pattern cladists more explicitly advocated such an approach when they rejected the use of evolutionary assumptions, arguing that we should classify on the "pure" observation of similarities and differences. So we have on one side, *empiricists*, who argue that classification should be independent of theory, and based on pure observation. And on the other side we have those who start with theory, and use it to guide observation and classification. Who is right? There is a long-standing debate in the philosophy of science about the relation between theory and observation, and perhaps we can better understand the debate about classification if we look at the larger historical context.

The Baconian Ideal

The empiricist ideal is widely associated with the views of English philosopher, statesman, lawyer, and writer Francis Bacon (1561–1626), whom we briefly encountered in Chapter 3. Bacon was a critic of both the old methods of learning and of old dogmas. He proposed that we start anew, first by replacing these old methods with new, and then setting aside all the old theories. Both projects were to be completed in his *Great Instauration*, a six-part masterwork. What we have of these new methods, though, are found largely in the only part of this work that was completed, the *New Organon* (*Novum Organum*) – a new "instrument" for rational inquiry. Here Bacon laid out his method. The fundamental idea behind this new method is that observation and experiment precedes theory. He rejected what might be called the "method of hypothesis" whereby we *start* with some theory, however we came to that theory, and then derive observable consequences or predictions that can either confirm or falsify that theory. Instead, on Bacon's approach, we arrive at theory only after, and through extensive observation and experiment.

Bacon rejected the "theory first" approach because he believed it to be a source of bias and error. As we saw in Chapter 3, Bacon thought of the mind as a mirror that reflects nature, but imperfectly so because of its distortions – the "idols of the mind." There we looked at one of these distortions, the "idols of the marketplace," based on language: how the words that we learn and use can confuse us about things in the world, taking things to be real that are not, and taking confused ideas about these things to be clear. But more to the point here are Bacon's "idols of the theatre," errors based on the acceptance of philosophical dogmas.

> Lastly, there are Idols which have immigrated into men's minds from the various dogmas of philosophies, and also from wrong laws of demonstration. These I call Idols of the Theatre, because in my judgment all the received systems are but so many stage plays, representing worlds of their own creation after an unreal and scenic fashion. Nor is it only of the systems now in vogue, or only of the ancient sects and philosophies, that I speak; for many more plays of the same kind may yet be composed and in like artificial manner set forth; see that errors the most widely different have nevertheless causes for the most part alike. Neither again do I mean this only of entire systems, but also of many principles and axioms in science, which by tradition, credulity, and negligence have come to be received. (Bacon 1960, Bk. I, XLIV)

Bacon has many targets here, from the grand systems of Plato, Aristotle, and the Neo-Platonists and Aristotelians who followed, to the Renaissance thinkers. But as the last sentence of this passage also indicates, he was also worried about the more narrow "principles and axioms" in science.

The new "inductive" method proposed by Bacon would bypass these theoretical biases by a careful foundation on observation. In his words, we should avoid all "anticipations of nature," hypotheses not inferred from sensory experience, and instead rely on the "interpretations of nature" that are well grounded on "the senses and particulars." He contrasted these two approaches, beginning with the anticipations:

> There are and can be only two ways of searching into and discovering truth. The one flies from the senses and particulars to the most general axioms, and from these principles, the truth of which it takes for settled and immovable, proceeds to judgment and to the discovery of middle axioms. And this way is now in fashion. The other derives axioms from the

> senses and particulars, rising by gradual and unbroken ascent, so that it arrives at the most general axioms last of all. This is the true way, but as yet untried. (Bacon 1960, Bk. I, XIX)

On both of these approaches we begin with observation, but the ascent to generalities in the anticipations bypasses what Bacon thought were crucial steps.

> Both ways set out from the sense and particulars, and rest in the highest generalities; but the difference between them is infinite. For the one just glances at experiments and particulars in passing, the other dwells duly and orderly among them. The one, again, begins at once by establishing certain abstract and useless generalities, the other rises by gradual steps to that which is prior and better known in the order of nature. (Bacon 1960, Bk. I, XXII)

In our anticipations of nature, we may start with something we see, feel, hear, or smell, and then adopt some general explanatory theory (perhaps some idol of the theatre) that then gets accepted and used to explain what we observe. These anticipations are powerful sources of bias, according to Bacon, because of their effect on the imagination:

> For the winning of assent, indeed, anticipations are far more powerful than interpretations, because being collected from a few instances, and those for the most part of familiar occurrence, they straightway touch the understanding and fill the imagination; whereas interpretations, on the other hand, being gathered here and there from very various and widely dispersed facts, cannot suddenly strike the understanding; and therefore they must needs, in respect of the opinions of the time, seem harsh and out of tune, much as the mysteries of faith do. (Bacon 1960, Bk. I, XXVIII)

It is human nature, an idol of the tribes, that humans "hurry toward certainty" based on imagination. So when presented with an anticipation or hypothesis, whether or not supported by the senses, the human mind naturally rushes to judgment and adopts that hypothesis.

Interpretations of nature avoid this effect on the imagination and rush to judgment through careful intermediate steps based on observation. After the first stage of pure observation of bodies and processes comes the stage of generalization and classification, whereby the patterns of qualities in things are summarized in three *tables of presentation* that group and organize things according to the presence, absence, or degree of qualities and properties. And from these carefully constructed tables of presentation we can infer the ultimate laws and principles of nature.

The *tables of existence and presence* group things according to common qualities or properties. Bacon illustrated this with his inquiry into the nature of heat, starting with a list of twenty-eight "instances meeting in the nature of heat." Among these instances are the sun's rays, reflected sun's rays, flaming meteors, lightning, any flame, and heated or boiling things. This is, in effect, a classification of things or processes that the senses tell us have heat. Then comes a presentation to the intellect of the instances "which are devoid of a given nature," a *table of divergence or related absences.* Bacon illustrated this by reference to those related cases where heat seems to be absent, such as the moon's rays and the sun's rays at higher elevations and polar regions (Bacon 1960, Bk. II, XI). The tables of presences and absences were to be completed in part by experiments that establish either the presence or the absence of particular qualities in cases not found in nature. Bacon suggested, for instance, the use of a "burning-glass" lens to determine whether the rays of the moon can be made to produce heat. Finally are the *tables of degrees or comparison.*

> Thirdly, we must make a presentation to the intellect of instances in which the nature under inquiry is found in different certain degrees, nor or less; which must be done by making a comparison either of its increase and decrease in the same subject, or of its amount in different subjects, as compared one with another. (Bacon 1960, Bk. II, XIII)

This is a method of determining the ways in which a thing can come to have different degrees of some quality. Bacon noted that the sun and planets, for instance, were thought to have more heat when in proximity to major fixed stars. Similarly some kinds of substances, such as burning tinder, have less heat, whereas other substances, burning coal and fired metals, have more.

The details of Bacon's method are beyond the scope of this book, but what is important is that Bacon is outlining a process of discovery, whereby we are to begin with observation and the senses, and only after that, through the tables of presentation, rise by inference to theories about laws, causes, and essential natures. Theories are *not* to be used to guide observation. Observation is therefore methodologically and logically prior to theorization. According to Bacon this method is the basis of an objective, unbiased science, because it allows us to avoid the *idols of the tribe* – rushing to conclusions, the *idols of the cave* – favoring our own personal preferences, and the *idols of the theatre* – interpreting observation on the basis of accepted dogmas or theories.

Bacon did not discuss biological classification in his *New Organon*, but we can see how his method could be applied. First, his tables of presentation result in classifications. The tables of presence group things according to the presence of a particular quality or property. The tables of absence exclude those things that do not have that quality or property. Bacon illustrated this process by applying it to things that have heat, but it could just as easily have been illustrated by reference to those things that have bony skeletons, or feathers. We could, for instance, write up a table of presence of feathers, and include all birds. Then the table of absence would exclude all those organisms with scales or fur. Second, these groupings – classifications, could then be used as part of further inquiry into the ultimate laws, causes, and essential natures. Why, for instance, do birds have feathers while bats do not? Or why do some birds have long elaborate tail feathers, and others do not?

The shortcomings of Bacon's method are apparent even with his chosen example of heat. He categorized things and phenomena into groups on the basis of whether they give the sensation of heat or not. But this is not how we now think about heat in terms of the kinetic energy of molecules. We now have a scientific theory that tells us what has heat, regardless of what the senses tell us. And this theory tells us that all things have heat, insofar as they are all composed of molecules with some nonzero quantity of kinetic energy. 'Heat' is now a theoretical term to be understood within a thermodynamic framework. But it isn't at all clear how a method that starts with observation and employs the tables of presentation *could* get us to such a theory. After all, we cannot observe molecules in motion, let alone something like average kinetic energy. Our modern science of thermodynamics does not seem very Baconian. But Bacon has nonetheless found followers over the centuries. Darwin himself claimed in his *Autobiography* to "have worked on true Baconian principles, and without any theory collected facts on a wholesale scale..." (F. Darwin 1887, 83).

We see a similarly empiricist approach in the method advocated by John Stuart Mill two centuries later, and that we briefly examined in Chapter 3. Mill explicitly advocated an approach to classification that is empiricist in the sense that it begins with observation, unbiased by theory. After recognizing that the mere use of general terms in a language generates classifications, and that the basis of a classification may depend on the purposes in classification, he nonetheless distinguished a natural classification from the more pragmatic "technical" classifications.

> The phrase Natural Classification seems most peculiarly appropriate to such arrangements as correspond, in the groups which they form, to the spontaneous tendencies of the mind, by placing together the objects most similar in their general aspect; in opposition to those technical systems which, arranging things according to their agreement in some circumstance arbitrarily selected, often throw into the same group objects which in the general aggregate of their properties present no resemblance, and into different and remote groups, others which have the closest similarity. It is one of the most valid recommendations of any classification to the character of a scientific one, that it shall be a natural classification in this sense also; for the test of its scientific character is the number and importance of its properties which can be asserted in common of all objects included in a group; and properties on which the general aspect of the things depends are, if only on that ground, important, as well as, in most cases, numerous. (Mill 1882, 872–873)

As indicated here, Mill did not think that all shared properties to be of equal value in generating a classification. Some would be more important than others. But which properties are most important is not determined by some theory. If so, a classification would be "technical." Instead what determines the significance of a property for generating a natural classification is the number of general propositions that can be made on the basis of that classification:

> The ends of scientific classification are best answered, when the objects are formed into groups respecting which a greater number of general propositions can be made, and those propositions more important, than could be made respecting any other groups into which the same things could be distributed. (Mill 1882, 871–872)

But how do we know which properties will support the most general propositions? According to Mill, causal properties, or "marks" of causal properties are best.

> The properties, therefore, according to which objects are classified, should, if possible, be those which are causes of many other properties; or, at any rate, which are sure marks of them. (Mill 1882, 872)

Mill is surely right here. If some property causes one or more other properties, then we can make generalizations on the basis of that property. If the hemoglobin in blood causes it to be red, for instance, then we can say that blood is red. And if it causes other properties as well, then we can assert propositions on the basis of these other properties. Causal properties make things alike or different in other ways. A consequence of this principle, according to Mill, is that "A natural arrangement, for example, of animals, must be founded in the main on their internal structure" (Mill 1882, 873). Internal structures have many effects that can provide support for many general propositions. On such a natural arrangement, whales are not fish because of the many internal differences with fish – even though there is a superficial resemblance – and for some technical and artificial purposes we might categorize them as fish (Mill 1882, 974–975).

But how do we know which properties are causal? We might answer that a *theory* could tell us which properties are causal. This is not Mill's approach though. In his *System of Logic*, he advocated an experimental method based on a *method of agreement* that looks a lot like Bacon's *tables of presence*, and a *method of difference* that looks a lot like Bacon's *tables of absence*:

> The simplest and most obvious modes of singling out from among the circumstances which precede or follow a phenomenon, those with which it is really connected by an invariable law, are two in number. One is, by comparing together different instances in which the phenomena occurs. The other is, by comparing instances in other respects similar in which it does not. These two methods may be respectively denominated, the Method of Agreement, and the Method of Difference. (Mill 1882, 478–479)

The idea here is that we just *look* at different instances of the same phenomena, and see how they are the same. And then we just *look* at instances that are similar in other ways, but don't exhibit the same phenomena, and see how they are different. If there is some property that is always present in a phenomenon and is always absent when the phenomenon is also absent, then that property must be necessary and sufficient for the effect. This is a method of discovery that does not seem to rely on any theoretical assumptions. Instead, *from* observation we infer causes and laws.

There are complexities in Mill's methods that we cannot address here, but the general idea is clear enough. Like Bacon, he argued that we should begin with pure observation and classify on that basis. Because such a classification is based solely on observation, this is an objective starting point and presumably a good foundation for knowledge. More recently John Wilkins and Malte Ebach have argued for a similarly empiricist approach. Although they seem to recognize multiple functions for classification, the empiricist function is the one they emphasize: "The role, or more correctly a role, of classification is to provide both the basis for discovery based upon patterns, which even if it is not induction is ampliative reasoning based upon as little theoretical foundation as possible" (Wilkins and Ebach 2014, 164). Here Wilkins and Ebach are clearly in this empiricist spirit that would have us start with "pure," atheoretical observation.

We saw this same philosophical commitment to empiricism in Chapter 4, in the views of the pheneticists. In their *Principles of Numerical Taxonomy*, Sokal and Sneath agreed with Mill that a classification is more natural to the degree that it supports a greater number of propositions (Sokal and Sneath 1963, 19). They argued that a classification based on overall similarity best accomplishes this. More narrow classifications based on some theoretical purpose will only narrowly and imperfectly serve that purpose. But as we saw in Chapter 4, the most explicit argument they made against theory was against a particular theory. Phylogenetic classifications, they argued, rely on circular reasoning. A classification representing phylogeny must be based on assumptions of homology – assumptions that particular shared traits are due to common ancestry. But these assumptions can themselves be confirmed only by a hypothesis of homology.

Like the pheneticists, pattern cladists argued that the purpose of classification is epistemic – to support propositions – which in turn help us discover new patterns.

> Why, then, do scientists concern themselves with constructing classifications? Perhaps because classifications serve not only to summarize information we already have, but also to predict new information we do

> not yet have. For example, if there is a set of organisms, all of which share properties (A, B, C, D, E) that no other organisms have, and we find another organism about which we know only that it has properties A and B, can we not predict that it will have properties C, D, and E as well? (Nelson and Platnick 1981, 9)

They thought this goal, as we saw in Chapter 4, to be best served by basing classification only on patterns of character distributions. Then phylogeny can be inferred from these patterns of character distributions. For these pattern cladists, *homologies* were not traits inherited from a common ancestor – they did not have a theoretical, evolutionary definition, but were simply characters held in common (Nelson and Platnick 1981, 208).

Observation and Theory

The intentions behind this empiricism are surely admirable. Progress in science requires that we set aside old inadequate or flawed theories in favor of new and better ones. This necessity would have been particularly compelling in Bacon's time. For nearly two millennia the Platonic and Aristotelian systems had been accepted dogma and had constrained thinking about the natural world. And Bacon was surely right about the pernicious influence of bias in his passages on the idols of the mind. Some biases arise from human perceptual tendencies, some from individual inclinations, and some from the effects of philosophical dogma. For an objective science based on facts about the world and not just our own flawed opinions, we need some way of correcting for or bypassing these biases. Plausibly the solution is to be found in sensory experience and observation.

The idea lurking here seems to go something like this. Human perception is flawed and limited (as Bacon realized), but in principle we can all look at the something and see it in the same way in some important sense. We can all look through a telescope or microscope, or at the outcome of an experiment and agree about what we see – even if there is variation in color perception or acuity of vision. Similarly we can all just look at an organism, or group of organisms, and see the same traits, characters, or properties. We can all see *that* an organism has a particular size or weight, a particular color, or a particular shape – even if our experience of these properties or traits may be different. And by extension we can all see the same patterns of similarities and differences in a group of organisms. We can see that one organism is bigger or smaller than another, that it has the same shape or color as another, or that one organism has wings and feathers, while another does not. Even with our individual idiosyncrasies in perceptual tendencies and capacities, we can agree about purely observable traits because our perception of the world is responsive to, and constrained by that world. Moreover, we can even agree about what we see in spite of the fact that we may be approaching the world with different theories. Both Linnaeus and Darwin could see that birds were feathered even though they had different theories about the origin of feathers. If so, then science will be objective in the sense that it is based on observable facts about the world, and not dependent on individual differences in perception and belief. And more to the point here, our *classifications* can also be objective and independent of human biases insofar as they are based on what is common in perception.

This is certainly an attractive ideal, and if realistic could ground an objective approach to science in general, and more specifically to biological classification. But there are some genuine problems. The first is how language influences our experience of the world. We began this book in Chapter 1 with the recognition that language and the words we learn both generate and influence a classification. When we learn the term 'Bengal tiger,' for instance, we learn two categories, a more general tiger category and a more specific subcategory that contains only the Bengal kind of tigers. One implication of this fact is that the categories we learn depend on the language and the words we learn. A child who hasn't learned the word 'tiger' will not have that category. This influence of language is not just limited to categories of organisms. It affects what traits, characters, or properties we categorize. When a child learns the word 'wings,' for instance, he or she acquires the category of wings. How he or she understands this category is partly a consequence of psychological biases for shape and function, as we saw in Chapter 1. Something may be taken to be a wing if it has some particular shape or functions in a particular way. Similarly, when we learn the words 'feather,' 'claw,' or 'eye' we learn a category for each of these things. And if we have learned these categories, we can use them in sensory experience of the world. We can see that an organism has wings, feathers, claws, and eyes. And we can see that another organism lacks these traits.

This language dependence may seem unproblematic. Perhaps many or even all languages have equivalent terms with associated categories for 'feather.' If so, then we could at least have intercultural agreement. But modern scientific classification relies on more problematic terms and categories. To classify organisms in modern systematics, a student might learn terms such as 'notochord,' 'amniote egg,' 'bilateral symmetry,' and 'three ear ossicles.' These are different from the folk terms 'wings' and 'eyes' in an important way. 'Notochord' and 'amniote egg' are technical terms in English, and may not have corresponding terms in all other languages. If so, then while a speaker of English may identify 'amniote egg' as a similarity between two organisms, a speaker of a language that doesn't contain that term may not do so. It is true that one might teach the term to the speaker who doesn't already have that term, and that would give this speaker a category in which to think about *amniote egg* as a similarity. This problem becomes even more apparent when we look at molecular traits. There are many sequences of DNA that have not been given names. But there are some that have been, such as the language gene *FOXP2*. Learning its name and associated category gives an additional trait that two individuals may or may not share. Someone who has not learned this term simply cannot compare two organisms on that basis. And to learn to apply this term would require learning some theories about DNA, genes, and language. There are certainly many people in all cultures all over the world who lack the linguistic and theoretical frameworks to identify *FOXP2* as a similarity or difference among organisms – even if confronted with some graphic representation of that gene. Observation alone doesn't give us the terms we might use to identify the similarities and differences that form the basis for classification.

The problem is this. Two people who have the same linguistic experience and knowledge may approach a group of organisms with the same categories and may then identify the same similarities and differences. But for two people who haven't learned the same set of general terms, there may be no single set of categories and similarities that they agree on. When they observe a set of organisms, they may identify a different set of similarities and differences – even though at some level they observe the same things. Observation, by itself, does not give the foundation for a single framework for determining similarities and differences. There is a linguistic component as well.

But why should we even learn the terms 'amniote egg,' 'notochord,' or '*FOXP2*'? The second problem with the empiricist approach is based on an insight of the twentieth century that has become known as the "theory-ladenness of observation." Each of these terms is a theoretical term, in that it is defined by reference to some theory or other. To learn the term 'amniote egg' and apply it to things in the world, for instance, one must have also learned a theory about development – what eggs are and how they function in development. Norwood Russell Hanson was an influential advocate of this understanding of observation. His crucial insight is that we *learn* to see and hear. A musician, for instance, learns to hear and a physicist learns to see.

> The infant and the layman can see: they are not blind. But they cannot see what the physicist sees; they are blind to what he sees. We may not hear that the oboe is out of tune, though this will be painfully obvious to the trained musician. (Who, incidentally, will not hear the tones and interpret them as being out of tune, but will simply hear the oboe to be out of tune.) (Hanson 1969, 17)

As suggested in the last, parenthetical sentence of this passage, this is not a compound process of seeing and then interpreting:

> We simply see what time is; the surgeon simply sees the wound to be septic; the physicist sees the X-ray tube's anode overheating. The elements of the visitor's visual field, though identical with those of the physicist, are not organized for him as for the physicist; the same lines, colors, shapes are apprehended by both, but not in the same way. (Hanson 1969, 17)

According to Hanson, we don't see things just by first recording the reception of photons on our retinas, and then interpreting the patterns of photons received as objects. Rather the experience is of seeing the world *as* composed of objects and processes. We don't see a pattern of colors for instance, and then interpret that pattern as a human, or a cat. We simply see something as a human or a cat.

This process of learning to see and hear should be familiar to anyone who listens to a musical composition. Only over time and repeated listening, and perhaps with guidance from a more knowledgeable listener, can one learn to hear the thematic structure. After doing so, the experience of listening to this particular music is very different than it was before. Similarly, we learn to see actions and processes in sports. We can see a "strike" in baseball, for instance, but only after we learn what a strike is, and what to look for. An important part of the education in any of the sciences is learning to observe. One must learn to see like an experimental physicist to become an experimental physicist. And one must learn to see like a botanist or geneticist to become a botanist or geneticist. To become a botanist, for instance, it would be necessary to learn to see *sepals, leaves, stamen,* and *pistils*. To do so, one must learn what these things are. And for the geneticist it is necessary to learn to see *chromosomes, genes, introns, exons,* and so on, which requires that the geneticist learn what these things are and how to observe them. To do this, a geneticist needs to learn a *theory* about genes.

Most relevant for purposes here, one must learn to see like a systematist to become a systematist. Precisely what that involves depends on what kinds of organisms and characters are being studied. A plant systematist will need to learn to see in a plant some of the same features a botanist sees – *sepals, leaves, stamens,* and *chloroplasts*, by learning how to apply the terms that refer to these features. For those who study vertebrates, there will be a different set of terms and characters that must be learned. That might involve learning the terms 'three ear ossicles,' 'incus,' 'stapes,' and 'malleus,' and what they refer to. Molecular systematists must learn a different set of terms that apply to molecular characters, and how to observe them. In each of these cases some theoretical background is required for observation. We can see these things only if we already have a theory that tells us what they are, and how to see them.

But even learning to see the things that these terms refer to does not automatically solve all problems. There is also a *character individuation* problem. Is *three ear ossicles* one character or three? It might be one character, but if one has also learned the names of the component ossicles – the *incus, stapes* and *malleus*, then perhaps there are three characters. Moreover, there might be features on each of these bones that can be named. If so, then each additional named feature might itself be a character. Similarly, we have the term 'feathered' that we can use in classification. But we might also make distinctions between the different kinds of feathers – tail feathers, wing feathers, and so on. Is a bird simply feathered? Or is it feathered in a particular way? How one identifies and individuates characters has implications for how many similarities and differences two organisms might possess. And as we saw in the discussion of phenetics in Chapter 4, this problem extends to the coding of traits. When we code *leaf length* we could code it in two states – short or long, three states – short, medium or long, or five states – very short, short, medium, long, or very long. On what *purely* observational grounds should we code leaf length in any of these ways? The bottom line is that any classification based on observed similarities and differences will depend on how characters are identified and individuated. But it isn't obvious that pure observation alone can tell us how to do this.

The Theory Theory

What all this suggests is that the Baconian ideal is unrealistic in that it doesn't seem to reflect the practice of modern science. Whether or not there is any pure, theory-neutral observation, it is clear that much of the observation in modern science is not. Observation in physics, chemistry, and biology seems to be systematically informed by theories about the world. There is a philosophical tradition that recognizes this. One of the seminal figures in this way of thinking is Immanuel Kant, whom we briefly encountered in Chapter 3. According to Kant, we don't get our concepts of *space*, *time*, *cause*, and *effect* from our sensory experience. Rather our experience presupposes these concepts. In his *Critique of Pure Reason*, he makes the case for our experience of things in space (emphasis added):

> Space is not an empirical concept which has been derived from outer experience. For in order that certain sensations be referred to something outside me (that is, to something in another region of space from that in which I find myself), and similarly in order that I may be able to represent them as outside and alongside one another, and accordingly as not only different but in different places, *the representation of space must be presupposed.* The representation of space cannot, therefore be empirically obtained from the relations of outer appearance. On the contrary, this outer experience is itself possible at all only through that representation. (Kant 1965, 68)

The important idea here is that we don't get our idea of space from the sensory experience of space, but from an antecedent representation. A careful analysis of Kant's views is beyond the scope of the project here, but we can see how this approach differs from the empiricist approach. On the empiricist approach, we would merely observe things in space, without any theory, and then formulate theories about space based on what we observe. But according to Kant, we instead experience space based on a prior concept that we have of space that guides our observation of things in space. In some sense this is a "theory" of space.

In the nineteenth century, the English polymath William Whewell adopted a view something like Kant's. Whewell began by claiming that science is an inductive process. But what Whewell meant by 'inductive' was not what Bacon or Mill meant. For Bacon and Mill, induction was generalization from observed instances to additional unobserved instances. We might, using Bacon's example, generalize from the properties of observed cases of heat to all cases of heat. Similarly, and according to Mill: "Induction, then, is that operation of the mind, by which we infer that what we know to be true in a particular case or cases, will be true in all cases which resemble the former in certain assignable respects"(Mill 1882, 354–355). We can see how induction in this sense might proceed from observation. We simply observe some property or quality in some instances and conclude that it is present in instances not yet observed.

Induction for Whewell was something very different. In his *Novum Organon Renovatum*, he contrasted the view associated with Bacon and Mill with his own:

> Induction is familiarly spoken of as the process by which collect a *General Proposition* from a number of *Particular Cases*: and it appears to be frequently imagined that the general proposition results from a mere juxta-position of the cases, or at most, from merely conjoining and extending them. But if we consider the process more closely, as exhibited in the cases lately spoken of, we shall perceive that this is an inadequate account of the matter. The particular facts are not merely brought together, but there is a New Element added to the combination by the very act of thought by which they are combined. There is a Conception of the mind introduced in the general proposition, which did not exist in any of the observed facts. (Whewell 1989, 140)

What is important in an induction, according to Whewell, is the application of a concept to the phenomena: "Thus in each inference made by Induction, there is introduced some General Conception, which is given, not by the phenomena, but by the mind" (Whewell 1989, 141). Whewell illustrated induction here by the application of the concept of *force* as the cause of change and motion. We observe the motion of bodies *as being* the products of forces – gravity, contact forces, and so on. When we see something fall, we see the motion as the product of a gravitational force. When we see one billiard ball hit another and cause it to move, we are seeing it within a theoretical framework of contact forces. And insofar as we always see this kind of motion in this way, the induction here is general.

In Whewell's view, as in Kant's, we don't simply observe the qualities and properties of things in the world and then generalize. We observe the world through the application of concepts to the world. If so, then we can see why Hanson's views, discussed earlier in this chapter, might be right. In science we need to *learn* how to see and hear. This learning involves the acquisition of the relevant concepts to apply to the world. We can apply this Whewellian notion of induction to biological classification. In our observation of organisms, we don't just see colors, shapes, and other properties, and classify. We apply concepts. We see an organism as having wings, because we have the concept of wings, and know how to apply it. Similarly, we see an organism as feathered, by applying the concept of feathers. This idea applies to many of the traits used to group and classify organisms: sepals, leaves, notochord, amniote egg, three ear ossicles, and more. To *see* a leaf, notochord, or amniote egg, we must have the concepts of leaf, notochord, and amniote egg. And to have these concepts we must learn them. This requires that we learn some theoretical background - a theory that tells us what counts as a leaf, notochord, or amniote egg.

This way of thinking about observation, as involving the application of a concept, has some obvious consequences for the idea that a classification can be purely empiricist and theory neutral. First, it is implausible that any classification can be completely atheoretical. The properties or traits that empiricists such as Buffon, Adanson, and de Jussieu, as well as the pheneticists and pattern cladists, use to group organisms into taxa must rely on concepts and theories. The use of *leaf length* by the pheneticists to group and classify, for instance, relies on the concept of *leaf*, which is dependent on some physiological theory that distinguishes leaves from other structures. Second, if Hanson is right that we must learn to observe, and that depends on the terms and theories that we learn, then there may be no single universal basis for *all* potential observers to classify things in the world. Two people that have different concepts and different background theories will not see things in precisely the same way. A New Guinea highlander who has never acquired the concept of *amniote egg*, its associated term, and the theory in which that concept appears, cannot apply it to an organism, and see that organism in precisely the same way as someone else who has acquired that concept and associated theory. Conversely, that highlander may have a concept and associated theory not possessed by those who see the *amniote egg*.

The Problem with Theories

For the reasons just considered, any methodological claim that classification should be completely theory free is highly problematic. But perhaps what is at issue is not this methodological claim about theory dependence in general, but about dependence on a particular theory. The pheneticists may have claimed to avoid theoretical bias in general, but their target was really a specific theory – evolutionary theory. As we saw in Chapter 4, they argued that an evolutionary classification based on phylogeny was circular because we could not determine which traits were due to common ancestry (homologies) and therefore useful for classification without already assuming phylogenetic relationships. We could, according to the pheneticists, determine phylogeny only on the basis of overall similarity. Here they were arguing against the use of *evolutionary* assumptions to classify. Similarly, the pattern cladists specifically rejected the use of *evolutionary* theory to classify. To be clear, it isn't that the pheneticists and the pattern cladists thought that evolution didn't occur. What they were rejecting were the specific ways evolutionary taxonomists and the phylogenetic cladists used evolutionary theory. To understand this use of evolutionary theory we need to return briefly to Darwin's evolutionary method of classification.

According to Darwin, as we saw in Chapter 3, organisms should be classified into species, and species should be classified into higher taxonomic levels on the basis of phylogeny or common ancestry. Darwin proposed a general method for determining phylogeny, based on character classification, the determination of whether a shared trait was homologous and due to common ancestry, or analogous (homoplasius) and due to parallel evolution or convergence by adaptation. As we saw in Chapter 3, Darwin's method of character classification was based on the following rule:

> *General Adaptation Rule*: A shared character trait that is likely to be an adaptation to a particular form of life is not likely to be homologous and is therefore irrelevant to classification.

He also gave guidance in the application of this rule, through a set of corollaries that were based on assumptions about the operation of natural selection: the *special habits corollary* that asserted the less any part of organization is concerned with special habits, the more important it is for classification; the *reproductive organs corollary*, that organs of reproduction are less likely to be special adaptations and are therefore more useful for classification; the *rudimentary organs corollary*, that atrophied instances of functional characters that have little current functional significance are unlikely to be adaptations, and are therefore useful to classification; and the *embryological corollary*, that embryological characters are unlikely to be adaptations and are therefore also useful. With each of these corollaries, Darwin was employing theoretical assumptions about how and when natural selection worked to determine whether a trait was likely to be an adaptation to a particular form of life or not, and consequently whether it was useful for classification.

The evolutionary taxonomists adopted some of Darwin's theoretical assumptions here. First they agreed with him that because classification should be on the basis of phylogeny, homologies and not analogies (homoplasies) were useful in classification, because the former but not the latter indicated phylogeny. They also agreed with him that functional analysis, to determine the operation of natural selection, would be useful in classifying characters as homologous. This, according to Simpson, was accomplished on the basis of assumptions about evolution: "evolutionary taxonomy must assume as a background the whole body of modern evolutionary theory, much of which is not directly taxonomic in nature but all of which has some bearing on taxonomy" (Simpson 1961, 82). Simpson proposed a series of rules intended to determine which characters are likely to be homologous. Although not all of his rules relied on assumptions about the operation of natural selection, several did. First is the *complexity rule*, that "Intricate adaptive complexes are unlikely to arise twice in exactly the same way, hence to be convergent in two occurrences, and the probability of homology is greater the more complicated the adaptation..." (Simpson 1961, 89). Second is the *special adaptation rule*: "...convergence is to be suspected when groups resemble each other only or most closely in some way specifically adaptive to a particular, shared ecology and are otherwise different" (Simpson 1961, 91). If shared traits are likely to be due to convergent evolution, then they are unlikely to be homologies and are therefore irrelevant to classification. Simpson concluded that any shared traits that were unlikely to be the products of adaptation by natural selection should be weighted more heavily.

The basic idea of functional analysis, as employed by both Darwin and Simpson, is that once we understand how characters function in an environment for a particular group of organisms, we can infer the operation of natural selection. And we can infer from that the probability that a shared trait is homologous and indicates phylogeny, or is homoplasious and does not. The full use of functional analysis, and the potential problems it poses, is apparent in a debate between the evolutionary taxonomist Walter Bock and the phylogenetic cladist Joel Cracraft, at a symposium on functional analysis at the 1979 meeting of the American Society of Zoologists. The papers presented at the symposium were then published two years later, in a 1981 issue of the journal *American Zoologist*. Bock argued for the approach to functional analysis associated with evolutionary taxonomy. While Cracraft accepted a phylogenetic basis of classification, he argued against this use of functional analysis in classifying characters and reconstructing phylogenies. Why he did so is instructive.

According to Bock phylogenetic hypotheses are tested against the "taxonomic properties" of characters – whether shared characters are due to convergent or parallel evolution (homoplasy), or common ancestry (homology), and if so, whether they are ancestral (symplesiomorphy) or more recent and derived (synapomorphy) (Bock 1981, 14). Bock argued that the establishment of these taxonomic properties of characters is the most important part of systematics because it constitutes the objective testing of a particular phylogenetic classification (Bock 1981, 5). The idea is that the hypothesis that two taxa are closely related is supported by similarities that seem to be recent homologies. The hypothesis is weakened by the presence of similarities that seem to be homoplasies – convergent or parallel evolution. According to Bock, these character classification hypotheses themselves are to be "tested" by reference to process hypotheses. He didn't say precisely how in the 1981 paper, but referred to a 1965 paper he had cowritten with Gerd van Wahlert.

In this earlier paper Bock and Wahlert distinguish *form*, *function*, *faculty*, and *biological use*. *Form* is simply the appearance or shape of feature:

> The form of a feature is simply its appearance, configuration, and so forth ... In morphology, the form would be the shape of the structure. In behavior, it would be the configuration of the display, including the involved structures, their movements, intensity and so forth. (Bock and van Wahlert 1965, 272–273)

The *function* of a feature is simply what it does on the basis of its form:

> Its function is its action, or simply how the feature works - as stemming from the physical and chemical properties of the form, a feature may have several functions that operate simultaneously or at different times. (Bock and van Wahlert 1965, 276)

A *faculty* is a form–function complex: "The faculty, comprising a form and a function of the feature, is what the feature is capable of doing in the life of the organism" (Bock and van Wahlert 1965, 276). A particular feature may have multiple forms, changing slowly as bones do in growth, or quickly as muscles do in contraction. But even with a single form, there may be multiple functions and faculties. A bone may function as a reservoir of mineral salts, by resistance to compression, or by resistance to shear. There will be a faculty associated with each of these functions of that particular form. Finally is *biological use*. A faculty may be either utilized or nonutilized. Those that get used acquire a biological role. A diving bird, for instance, may have a faculty for underwater "flying," but because of ecological circumstance never uses that faculty. The faculties that have biological roles are subject to natural selection on the basis of each role (Bock and van Wahlert 1965, 279). So once we know the biological roles of a feature we can then determine the likely operation of natural selection.

From what we have determined about biological roles, according to Bock and van Wahlert, we can determine first, whether a shared trait is a homology; second, the transformation sequence; and third, the direction of transformation. To accomplish the first task, they follow Darwin and Simpson, and argue that features that are likely to be adaptations to specific environments are unlikely to be homologies. The white coloration of arctic foxes, rabbits, and bears, for instance, is likely to be homoplasious based on what we know about the biological use of the white coloration both to avoid predation and to assist in predation. Functional analysis tells us that these traits are not homologies and are therefore irrelevant to classification.

Once it has been determined which characters are likely homologies, then these homologies are to be placed in a transformation series. Bock argued that this is accomplished only on the basis of known mechanisms of change. "Because transformation series describe the presumed evolutionary histories of features, it is not possible to establish or test them without a clear understanding of the mechanisms of adaptive evolutionary change of these features." These same mechanisms can help determine the "polarity" of a transformation sequence – the direction of change. Bock argued for what may be described as a polarity rule when he claimed that "evolutionary change will always be in the direction of the better adapted state" (Bock 1981, 16–17). On the basis of this rule, according to Bock, we can determine whether a homology is in an ancestral or a derived state. The better adapted state is the derived state. So if two groups of organisms share a character or trait that functional analysis tells us is derived and therefore of recent origin, we can infer that those two groups of organisms are closely related, sharing a recent common ancestor.

We cannot examine here the details of how this functional analysis is supposed to work (for that see Richards 2008), but what is important is the use of assumptions about evolutionary mechanisms to determine character classification. Cracraft criticized these assumptions in the 1979 symposium and the 1981 paper, where he argued first, that functional analysis is not possible; second that it is not useful; and third, that it is not necessary. Functional analysis is not possible because, according to Cracraft, it requires knowledge we do not have. We would need to know that a character has a genetic basis. We would need to know the genetic and phenotypic variability within the population it is found. Then we would need to know the relative fitness values for the various phenotypes. Only then could we say something about how natural selection would function within that population (Cracraft 1981, 31). In the absence of this information, according to Cracraft, we would simply be inventing functional and adaptationist stories to support whichever phylogenetic hypothesis we preferred (Cracraft 1981, 27–29).

Bock was also worried about the lack of this sort of information though:

> Study of the form and function of features in the laboratory, no matter how thoroughly done, is not a sufficient basis on which to determine adaptive significance. It is essential to do field work on the biological role of the features, on the environmental demands on the organism, and on the plausible selection forces. Only then, is it possible to determine the relationship between the feature and the selection force, to judge whether an adaptation exists, and to estimate its degree of goodness. The total work required to ascertain adaptations is far more difficult than usually believed, and errors can be made in the establishment of an adaptation even when the analysis is complete and thorough. (Bock 1981, 12)

Bock concluded: "Far fewer features have been shown to be adaptations in any detail and with any assurance than believed by most biologists" (Bock 1981, 12). The disagreement here about functional analysis does not seem to be about the availability of the required information. Both Cracraft and Bock agree that it is currently not available and difficult to get. But perhaps what is at issue is found in Cracraft's claim that functional analysis is irrelevant.

Cracraft didn't just say that functional analysis was difficult because of lack of the necessary information; he argued that "there is no way to evaluate the probability of a character transformation using functional data" (Cracraft 1981, 29). This suggests that *even if* we had the necessary information, we *still* couldn't classify characters and reconstruct transformation sequences. This is a very different claim and one that he did not elaborate on in this 1981 paper, except by making it clear that the role of natural selection is what is at issue:

> Natural selection cannot be claimed to be a universal causal mechanism for the origin and maintenance of phenotypic diversity and design. This is not to say, of course, that natural selection is not operating or that it is unimportant in evolution, only that it cannot be applied as a general law in evolutionary explanation. (Cracraft 1981, 33)

The problem here, according to Cracraft, is the assumption about the operation of natural selection beyond the population level: "...[I]t is clear that statements about natural selection should be restricted to the intrapopulation level of analysis; extending natural selection to specific and supraspecific levels ... further confuses the issue of process analysis" (Cracraft 1981, 45). This suggests that what bothered Cracraft were the assumptions about the operation of natural selection at the macroevolutionary level.

Lurking here is a dispute about the adequacy and accuracy of a particular "reductive" theory of evolution associated with the modern synthesis. Bock explicitly based his own views on this theory: "[M]y analysis is based on the acceptance of the reductionist model of macroevolutionary change postulated under the synthetic theory of evolution" (Bock 1981, 8). According to this "reductionist" model, long-term macroevolutionary change is just the sum of the changes within the relevant populations driven by natural selection. In his rejection of functional analysis above the population level, Cracraft was rejecting the assumption of this theory. He didn't have a lot to say about why in his 1981 paper, but he did in his 1980 book *Phylogenetic Patterns and the Evolutionary Process*, cowritten with Niles Eldredge, and published after the symposium but before the symposium paper. Here Cracraft (along with Eldredge) argued for a "decoupling" of long term macroevolutionary trends from the short-term microevolutionary trends within a species lineage. Microevolutionary trends can go one way, and the macroevolutionary trends can go another (Eldredge and Cracraft 1980, 285).

An example of this decoupling is found in one reconstruction of horse evolution. The fossil record suggests that the earliest horse species were smaller than the later more recent species, so that the long-term macroevolutionary trend is toward increasing size. This increasing size has sometimes been used to illustrate "Cope's rule," that natural selection generally results in an increase in size over time (Futuyama 1979, 139). The advantages of increased size seem obvious. Larger horses would be less vulnerable to predation and more resistant to environmental extremes. But horse evolution is not now seen as linear, with a single lineage producing larger horse species over time. Rather, horse evolution seems to be bushy, with multiple species lineages branching and producing new species or going extinct (McFadden 1986, 355). The problem is that within many of these branches, there may have been a microevolutionary trend toward decreasing size, even though the overall macroevolutionary trend has been toward increasing size (Jablonski 1997, 252). If so, the macroevolutionary trend has been decoupled from the microevolutionary trends. But how could this have happened?

In this book, Eldredge and Cracraft argued for a set of mechanisms that could explain such decoupling, based on speciation models, species selection, and species sorting. Speciation typically occurs in "peripheral isolates," according to Eldredge and Cracraft, when a small group of organisms gets reproductively isolated. Then that small group can undergo large change. But which group of organisms in a particular population forms a new species is determined by which gets isolated, not by which is the most fit relative to some environment. Moreover, which species persists over time might be a consequence, not of the individual fitness of its organisms, but of the species itself. Species selection, based on species-level traits rather than individual-level traits, would then determine macroevolutionary trends. Finally, which species survives might not be a product of selection forces at all, but a product of mere chance. One species might just have the bad luck to be in the wrong place at the wrong time and be decimated by a volcanic eruption, a meteor, a flood, or a fire, while another had good luck and survived. Again, which species survived was not a matter of the fitness of the individual organisms, but of random chance factors (Eldredge and Cracraft 1980, 277). Eldredge and Cracraft summarize these factors:

> At this juncture, we reiterate our acknowledgement of "random" factors: a new species may appear as the accidental by-product of change in the physical geography of the area ... A species might last longer than its sister in the next valley for purely accidental reasons, the converse of the observation that many extinctions are unlucky accidents. Entire ecosystems can be degraded relatively quickly ... In such cases, involving thousands of species in many unrelated clades, the environmental event (literally a catastrophe) occurs that is utterly accidental relative to individual species adaptations. (Eldredge and Cracraft 1980, 308)

This all implies, according to Eldredge and Cracraft, that we cannot extrapolate from the microevolutionary change that occurs within a population based on natural selection to the long-term macroevolutionary change that occurs over speciations and extinctions.

There are two models of macroevolution here, according to Eldredge and Cracraft. The "taxic model" is based on the formation of new species through peripheral isolation, species selection and species sorting. The "transformational model," on the other hand, is based on the idea that the processes operating within a species lineage – natural selection in particular, are also responsible for determining long-term trends. In the latter, but not the former, we can infer long-term macroevolutonary trends from short-term microevolutionary trends. This suggests that functional analysis based on the operation of natural selection might work well if the transformational model is correct, but it is much less likely to work if the taxic model is correct. In particular, if macroevolution is "transformational" we can infer the probability of a particular direction of change in a character – the polarity of a sequence, based on how that character functions and what natural selection would favor. But on the taxic model, the sequence of change in a character is determined at least partly by other random factors, geographic isolation and bad-luck extinctions. So it may be the case with horses that natural selection favored smaller size, but the "random" factors based on speciation, species selection, and species sorting produced an overall increase in size. If so, functional analysis, even if done correctly and with all the necessary information, would mislead us about the direction and polarity of change in size! This seems to be the reason that Eldredge and Cracraft claimed that "there is no way to evaluate the probability of a character transformation using functional data" (Cracraft 1981, 29).

There is much more to this debate about the nature of macroevolution, and I can hardly address it here. (See Richards 2008.) Nor do I want to come down on one side or other of the debate about macroevolution. But what is important is that the dispute over functional analysis is partly a dispute over two different sets of theoretical assumptions. Because Bock and Cracraft disagree about the nature of evolution, they disagree about how to determine phylogeny. The dispute is not about *whether* to use theory in classification, but about *which* theoretical assumptions to use.

The third criticism Eldredge and Cracraft had of functional analysis was that it was not necessary. Character classification could instead be accomplished on the grounds of the cladistic parsimony principle that asserts we should assume homology where possible. How this principle works was discussed in Chapter 4, so I won't review it here. It is worth noting, however, that the rejection of functional analysis and the transformation model of evolution is seemingly part of the "methodological" justification of that principle, although in a passage quoted in Chapter 4, Eldredge and Cracraft did not seem to acknowledge this fact:

> If one views the science of systematics as being subject to the same rules of inference as other branches of hypothetico-deductive science ... science must formulate a criterion by which to judge the relative merits of our close approximations (hypotheses). That criterion, in effect, is parsimony, and it specifies the most preferred hypothesis... (Eldredge and Cracraft 1980, 67)

If cladistic parsimony were just the application of a general methodological principle of science, it would not be based on any empirical assumptions about the world. But if the justification of parsimony is also based on the assumption of the taxic model of evolution over the transformational, then its justification is partly empirical – about the nature of the world and partly theoretical – about our best theories of the world.

Many of the early advocates of cladistics argued for the superiority of their methods on purely methodological grounds, often citing the philosopher of science Karl Popper. There are many problems with this use of Popper's philosophy, and I won't go into that debate here. (See Hull 1999 for a philosopher's take.) Rather I will just offer two speculative remarks. First, many times in science, disputes are ultimately about what empirical and theoretical assumptions are legitimate. But sometimes the disagreement about these assumptions cannot be resolved on empirical grounds. We cannot just look, for instance, and see that the taxic or transformational model is correct. Perhaps at least *sometimes* methodological claims might be made to avoid substantive assumptions about the world and on substantive grounds. We might argue for a particular method on purely methodological grounds, because we think the empirical or theoretical assumptions of another method are flawed, but cannot convincingly refute those assumptions. Second, there is a deeper philosophical issue here. Are there any purely methodological principles, or is methodology in science always dependent on some substantive assumption or other about the nature of things? It may be that there is no pure scientific methodology independent of what we think the world to be like.

This section is titled "the problem with theory." A clarification is now in order. What the discussion here seems to suggest is that a purely atheoretical classification is simply not possible. We think about the world in terms of concepts that have theoretically determined meanings. We cannot classify on the basis of amniote egg or leaf length, for instance, without knowing what they are, and that requires some theoretical foundation. If this is true, then our classifications will depend on how good are theories are. The "problem with theory" is that our classifications will only be as good or as bad as the theories on which they are based. But that doesn't mean that our classifications don't sometimes also serve our theories. That is the topic in the remainder of this chapter.

The Pragmatics of Classification

So far in this chapter we have two main conclusions. The first is that a classification cannot be completely independent of all theory if it is to use traits like *leaf length* and *amniote egg* to group and classify organisms. These traits seem to rely on some theoretical framework or other. The second is that when *theory* really is the issue, it may be the specific theoretical assumptions rather than theory dependence in general. But the relation between theory and classification may also be more complex than that. Much of the discussion in this book has been about the grouping of organisms into species, and species into higher-level Linnaean taxa. The discussion in this chapter has been about the relation between theory and this classification. But that is not the only way we can classify living things. We can also classify organisms on the basis of size - *microbe* and *macrobe*, trophic role - *predator* and *prey*, environment - *desert* versus *tropical forest*, physiological role - *parasite* and *host*, and so on. These classifications, unlike the Linnaean taxa, look more like the categories we find in the natural kinds framework. An organism is a member of the category *predator*, based not on its phylogeny, but on the presence of a particular behavioral trait - the tendency to feed on a particular set of organisms. An organism is a member of the category *desert flora*, on the basis of the environment in which it lives and thrives. And an organism is a member of the category *parasite* if it has a particular functional relation to another organism. In each case organisms are grouped and classified into *sets* on the basis of a particular property or relation. So even if there is some unique and special significance to the Linnaean classification of organisms and species, there are obviously other ways to classify and on other grounds.

In Chapters 5 and 7 we considered the pragmatic species pluralism advocated by Philip Kitcher (Kitcher 1992). According to this approach, there are multiple species concepts because there are many ways to group and divide organisms into sets on the basis of theoretical interests. Biologists might be interested in genetic or morphological similarities, the lineages represented in the fossil record, the processes that cause and maintain reproductive cohesion or isolation, or the functioning of natural selection. Hence they might classify on each of these grounds. This implies, according to Kitcher, that there is no single optimal way to think about and classify into species. Instead there are many legitimate ways, depending on theoretical interests. The obvious problem with this approach, as we saw in Chapter 5, is that these groupings are not all what we could plausibly take to be *species* groupings. Some groupings based on genetic similarities will correspond to species groupings, but others will not. But there is an insight here as well. Our classifications can *serve* theoretical interests as well as *reflect* theoretical assumptions. We might classify organisms into *parasite* and *host*, for instance, to serve theoretical purposes - to help us understand the nature of this relationship. That is the insight of pragmatic pluralism.

These classifications may be genuine natural kinds in some sense. *Predator* and *prey*, for instance, might be genuine natural kinds, in that they group and classify organisms into sets on the basis of theoretically important categories that are independent of human perception, interests, and preferences. These kinds might be classical in the sense that there is a set of well-defined necessary and sufficient conditions for being a *predator* or a *prey*. Or we might instead treat them as cluster kinds, where some subset of a larger set of conditions is sufficient. Similarly we might classify *parasite* and *host* on the basis of some physiological role, *herbivore* and *carnivore* on the basis of trophic behavior and physiology, and *marsupial* and *placental* on the basis of development and care for young. In each of these cases there are theoretical reasons to recognize these categories and classify on that basis.

In some cases these pragmatic categories might also correspond to the phylogenetic taxa of the evolutionary Linnaean framework. Because of common ancestry and selection pressures, there may be features of virtually all members of some taxon that is of some theoretical interest or other. We might, for instance, be interested in all members (or parts) of the species *Homo sapiens*, because of some unique "derived" features of the cerebral cortex that function in language. If so, then this pragmatic classification based on language use will roughly correspond to the phylogenetic taxon, in that they largely refer to the same group organisms. We might then be confused about the nature of the biological taxon, because there is pragmatic classification that corresponds to the evolutionary phylogenetic classifications. But at a more basic, fundamental level, these pragmatic classifications are very different from the evolutionary taxa of the Linnaean hierarchy. Instead of being the spatiotemporally located segments of populations lineages and branches on the evolutionary tree, they are sets of things where set inclusion is based on a set of properties or characters. It just so happens that all the organisms that have this set of properties are located in a particular time period and in a particular place. This correspondence of the pragmatic classification with the phylogenetic is perhaps part of the explanation for our tendencies to think about biological taxa as natural kinds – even though we also think about them in an inconsistent way as spatiotemporally located individuals.

There is no obvious reason we cannot be pragmatic pluralists, where our different ways of classifying organisms serve different theoretical and research interests. The species and higher-level taxa of the phylogenetic classifications might serve to represent the branching patterns of evolution that could then be used to develop, extend, and test theories in biogeography or character evolution. Once we have a well-supported tree, for instance, we might be able to evaluate Cope's rule that evolution tends toward increasing size. And each of the other ways that we group organisms into sets, based on a set of properties (whether essentialist or cluster) may also serve theoretical and research interests, helping to develop, extend, and test theories. By grouping organisms into *herbivore* and *carnivore*, or *host* and *parasite* we come to understand the processes at work in these relations. And we can modify our theories on the basis of what the classifications reveal. Similarly, for evolutionary reasons we might want to classify organisms into *sexual* and *asexual* to better study reproduction and understand the processes that always or sometimes operate in speciation. Perhaps these classifications are legitimate natural kinds. But there is no reason that we cannot also classify on conventional grounds. It is surely legitimate to classify and distinguish *pets* and *farm animals*, even though there may be no scientific theoretical significance with these terms. Rather, these classifications may be useful for commerce, law, or everyday life. And why shouldn't we classify organisms as *shellfish* for menu organization in restaurants, or as *vegetables* for nutritional purposes? This last category is useful so that we can tell children to eat their vegetables, even though there may be no legitimate scientific theoretical purpose. While neither of these categories corresponds to a legitimate biological taxon, they still serve some purpose or other. If these purposes are legitimate, then surely the classifications that serve them are legitimate as well.

But even so, there seems to be unique significance for the phylogenetic classification based on the branching and reticulating evolutionary tree. The evolutionary framework is not just any "theory" about the world. As Darwin recognized, and Dobzhansky articulated in his famous claim that "nothing makes sense in biology except in the light of evolution," evolutionary theory seems to play a foundational role in our thinking about living things. It is now standard to think about ecology, biogeography, physiology, genetics, population genetics, psychology, and social psychology within an evolutionary framework. This tendency is also found to an increasing degree in the human sciences and humanities, from anthropology and cognitive and evolutionary psychology to linguistics, evolutionary ethics, evolutionary epistemology, and evolutionary aesthetics. The fact that all of the characters, capacities, and properties that living things have are products of evolution is compelling reason to take an evolutionary, phylogenetic classification to be essential to our thinking about living things. This is so even if we have disagreements about precisely which processes produced these characters, capacities, and processes. We may disagree about the role of natural selection in evolutionary change, as Walter Bock and Joel Cracraft did, but that doesn't imply that the evolutionary framework is any less significant to our understanding of living things.

Conclusion: The Two Faces of Classification

This chapter has been about the relation between biological classification and theory. This relationship has sometimes been taken to be problematic based on the assumption that a classification based on a particular theory will be subjective and biased in favor of that theory. That classification will then trivially reflect whatever the theory assumes, regardless of the adequacy of the theory in representing things in the world. If so, how we think about things in the world just reflects the theories that we already have and perhaps prefer – even if there are better theories to be had. There may be some legitimacy to this worry, articulated most forcefully by Francis Bacon in his worries about the idols of the theatre, but the relationship between biological theories and classification is clearly more complicated than this simple model suggests.

What the analysis here seems to reveal is that, at the most basic level, biological classification has two faces, one turned toward the world and another toward theory. There is a face turned to the world in that classifications typically rely on features of the world that are independent of how we think about the world. In a phylogenetic classification, for instance, we classify on the basis of relations of ancestry and descent. That one organism is the offspring of another does not depend on how we think about the world. Or we might we might group organisms on the basis of their physical features or behavior, as we do when we think about organisms as having a particular size, an amniote egg, or as being a predator. Even a seemingly subjective classification such as *pleasing flowers* might still rely on physical proportions and a particular tendency toward the absorption and reflection of light. In each of these classifications, some objective human independent feature of the world plays a role in how living things are classified.

The other face of classification is turned toward theory. This may be a result of the fact that the classification is founded on some assumption of a theory, as we see in the modern phylogenetic, Linnaean system. This system assumes that organisms are part of a species lineage, they have common ancestry, and they originated in the branching speciation of evolution. Classification also faces theory because the terms and concepts on which a classification is based may be understood only in light of some theoretical framework. We saw this with terms such as 'amniote egg' that are defined in terms of some theory about development and its mechanisms. In the absence of any theory, it is unclear how we can even know what an "egg" is and how to identify one. The very generation of a classification based on a term and its associated concept requires something in addition to the human independent facts about the world. Whewell was surely right when he claimed that an important part of science is the application of a concept to the world. This application of a concept brings some theory along. Finally, classification faces theory in that it can function to further theoretical research goals. We might see this when we classify organisms as *sexual* or *asexual* to better understand and theorize about reproduction and its role in speciation.

If this is an accurate picture of classification, then the idea that we can have a completely theory neutral or theory independent classification is highly implausible. But the brute fact of theory dependence may not be what is problematic. The real problem may lie in *which* theoretical assumptions or *which* theoretical concepts we use, as we saw in the debate between Walter Bock and Joel Cracraft. Which *method* is right might ultimately depend on which *theory* is right. If so, then which classification is best might well depend on the relative merits of the theoretical and empirical assumptions behind each approach to classification. It then becomes imperative to have the right or the best theory. But this conclusion should hardly be surprising or problematic. Of course we need to have good theories about the world. Isn't that what science is all about, and the history of science reveals?

But what precisely is meant by the term 'theory' and its cognates? Nowhere in this chapter have I tried to work out a careful account of precisely what theories are, and what characterizes them. I suspect that because of the many different uses of 'theory' and its cognates, a universal theory of *theory* is not forthcoming. Perhaps we can just be satisfied with the observation that the learning of general terms often requires the acquisition of a concept that we apply to the world, and that we can often understand that concept only by reference to other concepts. So insofar as classification relies on the learning of general terms and concepts, it has a theoretical element, and in the context of a conceptual or theoretical framework. What makes something theoretical, including a biological classification, is simply the fact that a general term, a concept, a conceptual framework, and a set of related propositions are all involved. Only if a classification could be completely free of these things could it truly and fully be theory neutral. It isn't clear how this is possible.

9 The Essential Tension

The History of Biological Classification

One of the premises of this book is that we can better understand biological classification as it is currently practiced by looking at its history. In Chapter 2 we looked at the beginnings of modern biological classification in the use of the classificatory terms *eidos* and *genos* by Plato and Aristotle. Perhaps most instructive in this history are the different ways Aristotle used these terms. First was the narrow classificatory use in his method of difference, where an *eidos* was simply a subdivision of a *genos*, based on the presence of differentia. In the second usage, the *eidos* was the enmattered form, perpetuated in generation. Third was the development and articulation of a specific form out of a more generic matter. In the fifteen hundred years that followed, the commentators on Aristotle focused on the logical, classificatory use, in part because they read only Aristotle's logical works and were unaware of the other uses of these terms in his biological works. But early naturalists also used the Latin translations of these terms, 'species' and 'genus,' in an historical sense. Linnaeus, for instance, thought about species and higher taxa as lineages that extended back to the creation.

This historical understanding of biological taxa was fundamental to Darwin's evolutionary framework. Here species and biological taxa were branches on the evolutionary tree of life. Darwin's evolutionary approach was adopted first by the thinkers of the modern synthesis, most notably Ernst Mayr and G. G. Simpson, and then by Willi Hennig and the phylogenetic cladists. In spite of the criticism of this evolutionary approach by the pheneticists and pattern cladists, recent trends seem to favor an evolutionary, historical approach based on phylogenetic relationships. Species taxa are segments of population lineages and higher taxa are monophyletic branches on a branching and reticulating evolutionary tree. This historical way of conceiving biological taxa also seems to allow for a historical metaphysics. Taxa might plausibly be conceived as individuals with organisms as parts, rather than as sets of organisms where set inclusion is determined by possession of some set of properties or characters. Whether or not this conception of taxa as historical individuals becomes consensus, the evolutionary, historical way of thinking about biological taxa has clearly come to dominate biological classification.

A Puzzle

This historical way of thinking brings us back to a puzzle we introduced in the first chapter. There we looked at one standard way of thinking about scientific classifications, in terms of the natural kinds framework. On this framework, natural kinds are objective in the sense that they are independent of human beliefs, preferences, conventions, and biases. They are out there in nature, and are discovered, not invented. They "cut nature at its joints." So when we divide the substances of the world up into the elements, *hydrogen* and *oxygen*, and their compound into *water*, we have classified these substances into natural kinds. By contrast, *dollar bills*, *farm animals*, and *weeds* are conventional kinds, because they are the consequences of human conventions. And *blue foods*, *cars made on Monday*, and *things bigger than a breadbox* are arbitrary artificial kinds, because they don't even have a basis in human conventions.

The puzzle is that biological taxa seem to be real and objective just as natural kinds are real and objective. The existence and reality of *Homo sapiens* and *Aves* do not depend on mere human beliefs, conventions, or preferences, nor are they arbitrary like *blue foods*. So surely they must be natural kinds! But as we saw in Chapter 1, natural kinds are typically taken to be timeless and eternal because the set of properties that makes something a member of a natural kind are timeless and eternal. *Water* was H_2O in the past, it is in the present, and it will be in the future. As we saw in Chapter 5, attempts to understand species in terms of this natural kinds framework were fraught with all sorts of difficulties. It isn't obvious how to think about changing historical things within a framework that does not seem to allow change. Part of a solution to this puzzle (worked out in Chapter 6) may be to think about biological taxa as spatiotemporally restricted, historical individuals. Then species and higher-level taxa could exist and be real in a robust way, just as individual organisms exist and are real in a robust way, albeit to varying degrees.

But biological taxa still seem a lot like natural kinds in how they function epistemically and theoretically. Just as we can make inferences and generalize about all instances of oxygen, gold and water, we can make inferences and generalize about all (or most) members of a species, genus, or family. And just as the elements and the periodic table are theoretically significant, based on atomic number, and serve theoretical purposes, so are the species, genera, and families of biological classification, based on the processes that operate in evolution and that have created the evolutionary tree. So it appears that biological taxa seem natural kind–like, even though they cannot be natural kinds in the traditional sense, as sets of things, where set inclusion is determined by the presence of a timeless set of properties. How should we respond to this puzzle?

One response is to reform the natural kinds framework. Biological taxa may not be conceivable in terms of the essentialist framework, where set inclusion is determined by an unchanging and timeless set of necessary and sufficient conditions, because taxa change and are spatiotemporally situated. Nor may they be easily conceivable in terms of a cluster kind, where set inclusion is determined by the possession of some subset of properties or conditions, for the same basic reason. And as we saw in Chapter 6, although they might be conceived as historical kinds where set inclusion is based on historical relations, that way of thinking doesn't tell us how to segment lineages or branches into species or higher taxa. Mere historical relations may be necessary to group organisms into taxa, but aren't sufficient. Is there some way the natural kinds framework can accommodate the historical individuality of biological taxa?

Recently a naturalistic approach to natural kinds has been advanced, in the sense that we begin with the actual practice of science to see what counts as a natural kind. On this approach, briefly discussed in Chapter 8, natural kinds are simply those categories that are significant and essential to a particular science – whatever kinds of categories they might be. Muhammad Ali Khalidi argues for such an approach in his *Natural Categories and Human Kinds*:

> Natural kinds correspond to those categories that enable us to gain knowledge about reality. Since science is the enterprise dedicated to acquiring knowledge about the world, natural kinds are identified by the various branches of science. (Khalidi 2013, xi–xii)

This is also a deflationary account in that there is no single property that all natural kinds have in common. Each science might have different "kinds" of natural kinds, depending on which categories are necessary for gaining knowledge about reality. In chemistry, natural kinds may have one set of features, while in biology or psychology, natural kinds may have some other set of features. In psychology, for instance, human or mind independence cannot be a requirement for natural kindhood, because psychology is in part the study of the mind. The categories and divisions in psychology that help us understand minds and behavior will undoubtedly include mind-dependent kinds. The idea here is human minds and human behavior are just as much part of the natural world as are electrons, elements, and compounds, and should therefore be treated comparably (Khalidi 2013, 143).

Ali Khalidi argues that many natural kinds will be *etiological kinds*, based on causal history. He distinguishes "synchronic" causal properties that operate at a single time from "diachronic" properties – "phylogeny, descent or causal history." According to Al Khalidi, causal history based on evolution is most important in categorizing organisms, especially at higher taxonomic levels.

> The reason for the emphasis on causal history is not hard to ascertain in biology, since the theory of evolution is the central scientific theory in the biological sciences and evolutionary theory explains the diversity of life with reference to a history of natural selection. The difference between classifying by (synchronic) causal properties and by (diachronic) causal history does not show up very prominently in the classification of individual organisms into species since these two features tend to coincide at the level of individual organisms (those organisms with a shared causal history tend to share many causal properties in common and vice versa). But it does loom large when it comes to the classification of species and other taxa into higher taxa (genus, family, order and so on). Here, classification often sets aside similarities in causal properties (even nonsuperficial differences) in order to follow proximity of descent, or it tolerates significant differences in causal properties in order to track descent form common ancestor. (Khalidi 2013, 130–131)

So while natural kinds might have the traditional sets of essential properties in other domains, in biology they are historical entities connected by proximity of descent.

P. D. Magnus argues for a similarly pluralistic approach in his *Scientific Inquiry and Natural Kinds*:

> To put it loosely, the account of natural kinds which I defend maintains that a category of things or phenomena is a natural kind for a domain if it is indispensable for successful science of that domain. Scientific success involves making sense of the things or phenomena – both accurately predicting what they will do and explaining their features. The account conflicts with the tradition that associates natural kinds with fundamental and precise essences. (Magnus 2012, 2)

According to Magnus there are three constraints for natural kinds. They should support induction, figure in successful science, and be relative to domains of enquiry (Magnus 2012, 2). As with Ali Khalidi's approach, what counts as a natural kind may vary from one scientific context or domain of enquiry to another. Magnus specifically allows that species might be both spatiotemporally located individuals and natural kinds:

> The view that a species like *Homo sapiens* is an individual is often taken to be incompatible with the view that it is a natural kind, but note that this does not follow for the conception of natural kinds I have been advocating. If recognizing *Homo sapiens* is required for scientific success of the domain, then the species is a natural kind regardless of what it *is* at the level of deeper metaphysics. (Magnus 2012, 84)

If species (and perhaps higher level taxa as well) can be both natural kinds and individuals, then what has usually taken to be distinctive about natural kinds – that they are *sets* of things, may no longer be true.

This pragmatic approach to natural kinds has the advantage that it recognizes the differences in scientific domains. Chemistry and physics may have one way of dividing things in the world, but other sciences may do it other ways. If so, then a natural kinds framework that begins with kinds of particles, elements, or compounds may not reflect how other sciences make their divisions. Our natural kinds thinking might therefore be biased by the scientific domain that we begin with. Evolutionary biology, like geology, has a distinctively historical component. If biological taxa are branches on the evolutionary tree, then they must be historically situated. Similarly, geological strata are historically situated. The Navajo Sandstone and Kayenta Formation both have distinctive colors and features, but what makes a particular rock formation part of one of these strata, is partly *when* it was laid down. At some level, the pragmatic pluralism of Ali Khalidi and Magnus merely recognizes that not all sciences are the same. Moreover, biology and geology may not just be inferior versions of chemistry and physics.

Although this solves some problems with natural kinds thinking, there is a price to be paid. To give up the traditional notion of natural kinds means that they cannot play the philosophical role they have played so far, as David Hull explained more than twenty-five years ago:

> Redefining "natural kind" so that natural kinds are temporary and change into each other, sometimes gradually, sometimes abruptly, and still expecting them to play the same role that they always played in traditional philosophical systems, as as improbable as redefining "triangle" so that the sum of its internal angles is never 180° and expecting geometry to remain unchanged in the processes. Nothing stands in the way of a scientists redefining "natural kind," but in doing so, he has cut himself off from the philosophical literature in which this term occurs. (Hull 1989, 156)

What made the essentialist approach to natural kinds attractive, as traditionally conceived, is that it seemed to focus on what made exemplary natural kinds such as oxygen and water so significant. They are timeless and eternal and well defined in the sense that there is a small set of essential properties or conditions that make something oxygen or water. And there are a series of physical laws that apply to these kinds of natural kinds. For instance, the atomic number of the elements was used to place them on the periodic table, and this table formed the basis for the "periodic law," which asserts the chemical and physical properties of the elements recur periodically when the elements are placed in a table on the basis of atomic number. It is no accident that hydrogen and oxygen have the properties they have. Rather it is by the necessity of natural law. Given that hydrogen and oxygen have the atomic numbers they have, and that they are in their respective places in the periodic table, by scientific law and some sort of necessity they must have the suite of properties that they have.

But for historical individuals, natural laws do not seem applicable. As Hull points out: "If scientific laws have to be spatio-temporally unrestricted, and historical entities are defined in terms of spatio-temporal unity and continuity, then the two notions are clearly incompatible" (Hull 1989, 189). For historical individuals, the suite of properties they have seem to be contingent in a way that the properties of the elements are not. It is an accident of history that *Homo sapiens* turned out the way it did, and has the properties it has. If humans had evolved somewhere else and under different circumstances, then natural selection might well have produced different human characteristics. If the ancestors to modern humans had hybridized on a massive scale with other primate groups, they would have been different. Similarly, if a single organism had developed in a different environment, it might have different properties. So even if there are laws of development and natural selection, the nature of individuals seems contingent in ways that the nature of natural kinds, as traditionally conceived, is not. If biological taxa are not natural kinds in the traditional sense, then we can see why there are no laws of taxa. There is no law of *Homo sapiens* and there is no law of *Panthera tigris*. But if these two taxa are natural kinds, why are their natures contingent in a way that the natures of the elements are not?

The philosophical debate about the nature of scientific law is complex and has a long history and is beyond the scope of the discussion here. But one cannot just treat individuals as natural kinds without creating problems and puzzles for how we think about the divisions in the world, and their relation to scientific laws. But even more importantly, just *calling* biological taxa natural kinds doesn't solve the puzzle. What is truly puzzling is the fact that biological taxa are now usually taken to be historical, as segments of population lineages and branches on the evolutionary tree, yet we still seem to think of them as sets of things, where set inclusion is based on the possession of properties. To understand this, we need to return to a topic in the first chapter – how language generates classifications.

A Fundamental Tension

As we noted in the first chapter, we classify in part because we must classify. The mere learning of general terms in a language generates classifications. When we learn to use the term 'cat' we learn what that term applies to – things in the world that have particular properties or characteristics. And when we learn a general nominal modified by an adjective, such as 'black', we learn a hierarchy – *cat* as the general or generic category and *black cat* as a subcategory of cat. We learn to identify black cats by the color black and the properties we associate with cats – having a particular size, fur, claws, ears of a particular shape, and so on. Similarly when we learn to apply the Linnaean terms *Felis catus*, *Panthera tigris* or *Canis familiaris* we learn what properties are associated with each kind of thing. We don't learn to identify these biological taxa on the basis of their histories – the lineages in which they are a part.

Moreover, as we saw in Chapter 1, there seem to be clusters of biases in how we learn and generate these categories. In the first cluster, there is the whole object bias, in that when we first learn a term we tend to apply it to a whole object rather than a part or feature of that object. There is a principle of exclusivity where we assume there is only one name for each object. There is a principle of conventionality that assumes names don't change. In the second cluster, there is a principle of extendibility, where a word is taken to refer not just to single objects, but also to objects that are similar in various ways. Here there is a shape bias, where a term will be extended to objects that have a particular perceived shape. This explains how it is so easy to learn the term 'car' for a toy car and 'cat' for a small stuffed cat. But terms can also be extended on the basis of function. The term 'chair' can be applied to things that have different shapes if they have the same function. In the third cluster is a series of biases related to the development of hierarchies. When children learn different syntactic functions – to distinguish adjectives from nouns, they also learn subcategories. And some of these levels of categories are also treated as more basic. The basic level seems to be the most inclusive level where members possess numerous common attributes. They have the most similarity. This is the first, most easily learned, and most likely named level. The fourth cluster of biases is to treat these categories as having essences – some essential property or set of properties that make a thing what it is, and is required to be that kind of thing.

What is important here is that these biases seem to lead us to think about things in the world as belonging to sets on the basis of particular properties. In effect, we have an innate psychological *set bias*. When learning a general term we tend to think of our categories as being sets of things, where set inclusion is based on shape and functional properties and these properties constitute the essence of the things in the set. When we learn the term *Panthera tigris* we naturally learn it as applying to a set of things with particular properties. But when we learn the evolutionary, phylogenetic classification of living things in the modern Linnaean framework, we learn to classify instead on the basis of location in an historical lineage and the tree of life. Our psychological tendencies and biases are then leading us astray. There is then a fundamental tension between our psychological biases and the theory that governs biological classification.

This tension has played out in the history of classification from Aristotle to the present. We see it in Aristotle's two uses of the terms 'eidos' and 'genos.' In the sets usage, an eidos is merely a subset of a more general *genos* based on some differentia. But in the usage where an *eidos* is an enmattered form perpetuated in generation, the thinking is historical much like modern evolutionary thinking. Aristotle was not an evolutionist, but he did recognize that biological taxa also have an historical component. This psychological bias towards sets dominated in the centuries that followed in the thinking of those who read only Aristotle's work, and were most interested in the linguistic problem of universals. But the tension between the historical and the psychological sets bias returned in the early naturalists like Linnaeus, who thought about taxa as historical lineages. Recall that Linnaeus used sexual traits to classify but also recognized that this was an artificial method. In part this was because just a few traits could not represent the divine ideas that governed creation and persisted in the lineages descended from the original created organisms. Similarly, Buffon thought that classification should be based on all traits, yet he nonetheless recognized that species were historical things – lineages of organisms. And in the disputes between the evolutionary taxonomists and pheneticists we also saw this tension. Certainly biological taxa share similarities, as the pheneticists emphasized, but they are also branches on the evolutionary tree, as the evolutionary taxonomists argued. This tension was also found in the disputes between the phylogenetic and pattern cladists, the former adopting the theoretical historical framework, the latter following the psychological sets bias.

Until human psychology changes, and there is little reason to believe it will soon, and as long as we learn and apply general terms, this tension will remain. We can learn the theoretical framework of modern evolutionary biological classification, but we cannot avoid our psychological tendencies in learning the general terms that we apply to biological taxa. We cannot avoid thinking about all the kinds of organisms we see in the world as timeless sets of things, where set inclusion is determined by a set of properties, even though we also believe these kinds to be historical and branches on an evolutionary tree. If so, then the practice of biological classification will continue to be fraught with the problems that arise from this fundamental tension.

References

Agapow, P. M., O. R. P. Bininda-Edmunds, K. A. Crandall, J. L. Gittleman, G. M. Mace, J. C. Marshall, and A. Purvis. 2004. "The Impact of Species Concept on Biodiversity Studies." *The Quarterly Review of Biology*, 79(2): 161–179.

Amundson, R. 2005. *The Changing Role of the Embryo in Evolutionary Thought*. Cambridge: Cambridge University Press.

Andam, C. P. D. Williams, and J. P. Gogarten. 2010. "Natural Taxonomy in Light of Horizontal Gene Transfer." *Biology and Philosophy*, 25: 589–602.

Aristotle. 1995. *The Complete Works of Aristotle*, Vols. I and II, ed. J. Barnes. Princeton, NJ: Princeton University Press.

Atran, S. 1999. "Itzaj Maya Folkbiological Taxonomy," in *Folkbiology*, ed. D. L. Medin and S. Atran, 119–203. Cambridge, MA: MIT Press.

Bacon, F. 1960. *The New Organon and Related Writings*, ed. F. H. Anderson. Indianapolis, IN: Bobbs-Merrill.

Balme, D. M. 1987. "Aristotle's Use of Division and Differentiae," in *Philosophical Issues in Aristotle's Biology*, ed. A. Gotthelf and J. G. Lennox, 69–89. Cambridge: Cambridge University Press.

Bapteste, E., and R. Burian. 2010. "On the Need for Integrative Phylogenomics, and Some Steps toward Its Creation." *Biology and Philosophy*, 25: 711–736.

Barnes, J. 1995. "Life and Work," in *The Cambridge Companion to Aristotle*, ed. J. Barnes, 1–26. Cambridge: Cambridge University Press.

Baum, D. A., and S. D. Smith. 2012. *Tree Thinking: An Introduction to Phylogenetic Biology*. Greenwood Village, CO: Roberts & Company.

Baum, D. A., S. D. Smith, and S. S. S. Donovan. 2005. "The Tree-Thinking Challenge." *Science*, 310(5750): 979–980.

Berlin, B. 1992. *Ethnobiological Classification: Principles of Categorization of Plants and Animals in Traditional Societies*. Princeton, NJ: Princeton University Press.

1999. "How a Folkbotanical System Can Be Both Natural and Comprehensive: One Maya Indian's View of the Plant World," in *Folkbiology*, ed. D. L. Medin and S. Atran, 71–89. Cambridge: MA: MIT Press.

Bird, A., and E. Tobin. 2008. "Natural Kinds," in *The Stanford Encyclopedia of Philosophy* (Summer 2009 edition), ed. E. N. Zalta. Retrieved from: http://plato.stanford.edu/archives/sum2009/entries/natural-kinds/

Blackburn, S. 2002. "Metaphysics," in *The Blackwell Companion to Philosophy*, ed. N. Bunnin and E. P. Tsui-James, 61–89. Malden, MA: Blackwell.

Blumenthal, H. J. 1996. *Aristotle and Neoplatonism in Late Antiquity: Interpretations of De Anima*. Ithaca, NY: Cornell University Press.

Bock, W., and G. v. Wahlert. 1965. "Adaptation and the Form-Function Complex." *Evolution*, 19(3): 269–299.

1981. "Functional-Adaptive Analysis in Evolutionary Classification." *American Zoologist*, 21: 5–20.

Bouchard, F. 2013. "What Is a Symbiotic Individual and How Do You Measure Its Fitness?" in *From Groups to Individuals: Evolution and Emerging Individuality*, ed. F. Bouchard and P. Huneman, 243–264. Cambridge, MA: MIT Press.

Boyd, R. 1999. "Homeostasis, Species, and Higher Taxa," in *Species: New Interdisciplinary Essays*, ed. R. A. Wilson, 141–185. Cambridge MA: MIT Press.

Brogaard, B. 2004. "Species as Individuals." *Biology and Philosophy*, 19: 223–242.

Brown, S. F. 1999. "Realism versus Nominalism," in *The Columbia History of Western Philosophy*, ed. R. Popkin. New York: Columbia University Press.

Bulmer, R. N. H. 1967. "Why Is the Cassowary Not a Bird? A Problem of Zoological Taxonomy among the Karam of the New Guinea Highlands." *Man*, 2: 5–25.

Capone, N. C., W. O. Haynes, and K. Grohne-Riley. 2010. "Early Semantic Development: The Developing Lexicon," in *Language Development: Foundations, Processes, and Clinical Applications*, ed. B. B. Shulman and N. C. Capone. Burlington, MA: Bartlett.

Clarke, E., and S. Okasha, S. 2013. "Species and Organisms: What Are the Problems?" in *From Groups to Individuals: Evolution and Emerging Individuality*, ed. F. Bouchard and P. Huneman, 197–223. Cambridge, MA: MIT Press.

Cracraft, J. 1981. "The Use of Functional and Adaptive Criteria in Phylogenetic Systematics." *American Zoologist*, 21: 21–36.

2000. "Species Concepts in Theoretical and Applied Biology: A Systematic Debate with Consequences," in *Species Concepts and Phylogenetic Theory*, ed. Q. D. Wheeler, and R. Meier, 3–14. New York: Columbia University Press.

Crisp, A., C. Boschetti, A. Tunnacliffe, and G. Micklem. 2015. "Expression of Multiple Horizontally Acquired Genes Is a Hallmark of Both Vertebrate and Invertebrate Genomes." *Genome Biology*, 16: 50, DOI 10.1186/s13059-015-0607-3.

Dagan, T., Y. Artzy-Randrup, and W. Martin, 2008. "Modular Networks and Cumulative Impact of Lateral Transfer in Prokaryote Genome

Evolution." *Proceedings of the National Academy of Sciences of the USA*, 105(29): 10039–10044.

Darwin, C. 1859. *On the Origin of Species*, at *The Complete Works of Charles Darwin Online*. Retrieved from: http://darwin-online.org.uk/content/frameset?viewtype=text&itemID=F339.1&keywords=cirripedia&pageseq=1

Darwin, F. 1887. *The Life and Letters of Charles Darwin*. New York: D. Appleton and Company.

Dennett, D. 1995. *Darwin's Dangerous Idea*. New York: Simon and Schuster.

Depew, D., and B. H. Weber 1995. *Darwinism Evolving: Systems Dynamics and the Genealogy of Natural Selection*. Cambridge, MA: MIT Press.

de Queiroz, K. 1999. "The General Lineage Concept of Species and the Defining Properties of the Species Category," in *Species: New Interdisciplinary Essays*, ed. R. A. Wilson, 49–89. Cambridge, MA: MIT Press.

2005. "Ernst Mayr and the Modern Concept of Species." *Proceedings of the National Academy of Sciences of the USA*, 102(1): 6600–6607.

de Queiroz, K., and P. D. Cantino. 2001. "Taxon Names, not Taxa Are Defined." *Taxon*, 50(3): 821–824.

de Queiroz, K., and J. Gauthier. 1992. "Phylogenetic Taxonomy." *Annual Review of Ecology and Systematics*, 23: 449–480.

1994. "Toward a Phylogenetic System of Biological Nomenclature." *Trends in Ecology and Evolution*, 9: 27–31.

Devitt, M. 2008. "Resurrecting Biological Essentialism." *Philosophy of Science*, 75(3): 344–382.

Diamond, J., and K. D. Bishop, 1999. "Ethno-ornithology and the Ketengban People, Indonesian New Guinea," in *Folkbiology*, ed. L. Medin and S. Atran. 17–45. Cambridge: MA: MIT Press.

Dobzhansky, T. 1937. *Genetics and the Origin of Species*. New York: Columbia University Press.

1964. "Biology, Molecular and Organismic." *American Zoologist*, 4(4): 443–452.

Doolittle, W. F. 2010. "The Attempt on the Life of the Tree of Life: Science, Philosophy and Politics." *Biology and Philosophy*, 25: 455–473.

Dupré, J. 1993. *The Disorder of Things: Metaphysical Foundations of the Disunity of Science*. Cambridge, MA: Harvard University Press.

Ebbesen, S. 1990a. "Boethius as an Aristotelian Commentator," in *Aristotle Transformed*, ed. R. Sorabji, 373–391. Ithaca, NY: Cornell University Press.

1990b. "Porphyry's Legacy to Logic: A Reconstruction," in *Aristotle Transformed*, ed. R. Sorabji, 141–179. Ithaca, NY: Cornell University Press.

Eddy, J. H. Jr. 1994. "Buffon's Histoire Naturelle: History? A Critique of Recent Interpretations." *ISIS*, 85(40): 644–661.

Eldredge, N., and J. Cracraft. 1980. *Phylogenetic Patterns and the Evolutionary Process*. New York: Columbia University Press.

Ellstrand, N. C., R. Whitkus, and L. H. Rieseberg. 1996. "Distribution of Spontaneous Plant Hybrid." *Proceedings of the National Academy of Sciences of the USA*, 93: 5090–5093.

Ereshefsky, M. 1992. "Species, Higher Taxa, and the Units of Evolution," in *The Units of Evolution: Essays on the Nature of Species*, ed. M. Ereshefsky. Cambridge, MA: Bradford Books.

1998. "Species Pluralism and Anti-Realism." *Philosophy of Science*, 65: 103–120.

2001. *The Poverty of the Linnaean Hierarchy: A Philosophical Study of Biological Taxonomy*. Cambridge: Cambridge University Press.

2010. "Microbiology and the Species Problem." *Biology and Philosophy*, 25: 553–568.

Eriksson, G. 1983. "Linnaeus the Botanist," in *Linnaeus the Man and His Work*, ed. T. Frangsmyr. Berkeley: University of California Press.

Falcon, A. 2005. "Commentators on Aristotle," in *The Stanford Encyclopedia of Philosophy*, ed. Edward N. Zalta. Retrieved from http://plato.stanford.edu/archives/fall2005/entries/aristotle-commentators/

Felsenstein, J. 1982. "Numerical Methods for Inferring Evolutionary Trees." *Quarterly Review of Biology*, 57(4): 379–404.

1983. "Methods for Inferring Phylogenies." in *Numerical Taxonomy*, ed. Joseph Felstein. Berlin: Springer-Verlag.

Franklin-Hall, L. R. 2010. "Trashing Life's Tree." *Biology and Philosophy*, 25: 689–709.

Futuyama D. 1979. *Evolutionary Biology*. Sunderland, MA: Sinauer Associates.

Gelman, S. A. 2003. *The Essential Child: Origins of Essentialism in Everyday Thought*. Oxford: Oxford University Press.

Ghiselin, M. T. 1969. *The Triumph of the Darwinian Method*. Chicago: University of Chicago Press.

1997. *Metaphysics and the Origin of Species*. Albany: State University of New York Press.

Griffiths, P. 1999. "Squaring the Circle, Natural Kinds with Historical Essences," in *Species, New Interdisciplinary Essays*, ed. R. A. Wilson, 209–228. Boston, MA: MIT Press.

Haber, M. H. 2012. "Multilevel Lineages and Multidimensional Trees: The Levels of Lineage and Phylogeny Reconstruction." *Philosophy of Science*, 79: 609–623.

2013. "Colonies Are Individuals," in *From Groups to Individuals: Evolution and Emerging Individuality*, ed. F. Bouchard and P. Huneman, 195–217. Cambridge, MA: MIT Press.

Haber, M. H., and A. Hamilton. 2005. "Coherence, Consistency, and Cohesion: Clade selection in Okasha and Beyond." *Philosophy of Science*, 72: 1026–1040.

Hanson, N. R. 1969. *Patterns of Discovery*. Cambridge: Cambridge University Press.

Hennig, W. 1966. *Phylogenetic Systematics*, trans. D. D. Davis and R. Zangerl. Urbana: University of Illinois Press.

Hey, J. 2001. *Genes, Categories and Species: The Evolutionary and Cognitive Causes of the Species Problem*. Oxford: Oxford University Press.

Huerta-Sánchez, E., X. Jin, Asan, Z. Bianba, B. M. Peter, N. Vinchenbosch, Y. Liang, X. Yi, M. He, M. Somel, P. Ni, B. Wang, X. Ou, Huasang, J. Luosang, Z. X. P. Cuo, Kui Li, G. Gao, Y. Yin, W. Wang, X. Zhang, X. Xu, H. Yang, Y. Li, J. Wang, J. Wang, and R. Nielsen. 2014. "Altitude Adaptations in Tibetans Caused by Introgression of Denisovan-like DNA." *Nature*, 512: 194–197.

Hull, D. L. 1989. *The Metaphysics of Evolution*. Albany: State University of New York Press.

1990. *Science as Process: An Evolutionary Account of the Social and Conceptual Development of Science*. Chicago: University of Chicago Press.

1992. "A Matter of Individuality," in *The Units of Evolution: Essays on the Nature of Species*, ed. M. Ereshefsky, 293–316. Cambridge, MA: Bradford Books

1998. "Introduction to Part V," in *The Philosophy of Biology*, ed. D. L. Hull and M. Ruse. Oxford: Oxford University Press.

1999. "The Use and Abuse of Sir Karl Popper." *Biology and Philosophy*, 14: 481–504.

Hunn, E. 1982. "The Utilitarian Factor in Folk Biological Classification." *American Anthropologist*, 84(4): 830–847.

Huxley, J. S. 1940. "Introductory: Towards the New Systematics," in *The New Systematics*, ed. Julian Huxley, 1–46. London: Oxford University Press.

Hyman, A., and J. J. Walsh (1983) *Philosophy in the Middle Ages*. Indianapolis, IN: Hackett.

Jablonski, D. 1997. "Body Size Evolution in Cretaceous Molluscs and the Status of Cope's Rule." *Nature*, 385: 250–252.

2008. "Species Selection: Theory and Data." *Annual Review of Ecology, Evolution and Systematics*, 39: 501–524.

Jones, W. T. 1969. *The Medieval Mind*. San Diego: Harcourt Brace Jovanovich.

Kant, I, 1965. *Critique of Pure Reason*, trans. N. K. Smith. New York: St. Martin's Press.

Khalidi, M. Ali 2013. *Natural Categories and Human Kinds: Classification in the Natural and Social Sciences*. Cambridge: Cambridge University Press.

Kitcher, P. 1992. "Species," *The Units of Evolution: Essays on the Nature of Species*, ed. M. Ereshefsky, 317–341. Cambridge, MA: Bradford Books.

Kitts, D. R., and D. J. Kitts. 1979. "Biological Species as Natural Kinds." *Philosophy of Science*, 46(4): 613–622.

Klima, G. 2003. "Natures: The Problem of Universals." *Cambridge Companion to Medieval Philosophy*, ed. A. S. McGrade, 196–206. Cambridge University Press.

Kripke, S. 1972. *Naming and Necessity*. Cambridge, MA: Harvard University Press.

LaPorte, J. 2004. *Natural Kinds and Conceptual Change*. Cambridge: Cambridge University Press.

Larson, J. L. 1968. "The Species Concept of Linnaeus." *Isis*, 59(3): 291–299.

1971. *Reason and Experience: The Representation of Natural Order in the Work of Carl von Linne*. Berkeley: University of California Press.

Lecointre, G., and H. L. Guyader. 2006. *The Tree of Life: A Phylogenetic Classification*. Cambridge, MA: Harvard University Press.

Leibniz, G. 1996. *New Essays on Human Understanding*, transl. and ed. P. Remnant and J. Bennett. Cambridge: Cambridge University Press.

Lennox, J. 1980. "Aristotle on Genera, Species and the More and the Less." *Journal of the History of Biology*, 13: 321–346.

1987. "Kinds, Forms of Kinds and the More and Less in Aristotle's Biology," in *Philosophical Issues in Aristotle's Biology*, ed. A. Gotthelf and J. G. Lennox, 339–359. Cambridge: Cambridge University Press.

2006. "Aristotle's Biology," in *The Stanford Encyclopedia of Philosophy*, ed. E. N. Zalta. Retrieved from http://plato.stanford.edu/archives/fall2006/entries/aristotle-biology/

Lindroth, S. 1983. "The Two Faces of Linnaeus," in *Linnaeus: The Man and His Work*, ed. T. Frangsmyr, 1–62. Berkeley: University of California Press.

Linnaeus, C. 1964. *Systema Naturae, Facsimile of the First Edition*, ed. M. J. S. Engel-Ledeboer and H. Engel. Nieuwkoop, Netherlands: B. De Graaf.

Locke, J. 1975. *An Essay Concerning Human Understanding*. Oxford: Clarendon Press.

Lovejoy, A. O. 1968. "Buffon and the Problem of Species," in *Forerunners of Darwin*, 84–113. Baltimore, MD: Johns Hopkins University Press.

MacLauren, J., and K. Sterelny. 2008. *What Is Biodiversity?* The University of Chicago Press.

Maddison, D. R., K-S. Schulz, and W. P. Maddison 2007. "The Tree of Life Web Project." *Zootaxa*, 1668: 19–40.

Madigan, A. S. J. 1994. "Alexander on Aristotle's Species and Genera as Principles," in *Aristotle in Late Antiquity*, ed. L. Schrenk, 76–91. Washington, DC: The Catholic University of America Press.

Magnus, P. D. 2012. *Scientific Inquiry and Natural Kinds: From Planets to Mallards*. London: Palgrave Macmillan.

Maier, D. S. 2013. *What's So Good about Biodiversity*. Dordrecht, the Netherlands: Springer.

Mallet, J. 2005. "Hybridization as an Invasion of the Genome." *Trends in Ecology and Evolution*, 20(5): 229–237.

2013. "Darwin and Species," in *The Cambridge Encyclopedia of Darwin and Evolutionary Thought*, ed. Michael Ruse, 109–115. Cambridge: Cambridge University Press.

Marenbon, J. 2005. "Anicius Manlius Severinus Boethius," in *The Stanford Encyclopedia of Philosophy* (Fall 2008 Edition), ed. Edward N. Zalta. Retrieved from http://plato.stanford.edu/archives/fall2008/entries/boethius/

Markman, E. 1989, *Categorization and Naming in Children: Problems of Induction*. Cambridge, MA: MIT Press.

Marrone, S. P. 2003. "Medieval Philosophy in Context," in *Cambridge Companion to Medieval Philosophy*, ed. A. S. McGrade, 10–50. Cambridge: Cambridge University Press.

Mason, A. S. 2010. *Plato*. Berkeley: University of California Press.

Mayden, R. L. 1997. "A Hierarchy of Species Concepts: The Denouement in the Saga of the Species Problem," in *Species: the Units of Biodiversity*, ed. M. F. Claridge, H. A. Dawah, and M. R. Wilson, 381–424. London: Chapman and Hall.

Mayr, E. 1942. *Systematics and the Origin of Species*. New York: Columbia University Press.

1982. *The Growth of Biological Thought*. Cambridge, MA: Belknap Press.

Mayr, E., and P. D. Ashlock. 1991. *Principles of Systematic Zoology*. New York: McGraw-Hill.

McFadden, B. 1986. "Fossil Horses from "Eohippus" (Hyracotherium) to Equus" Scaling, Cope's Law and the Evolution of Body Size." *Paleobiology*, 12(4): 355–369.

McOuat, G. R. 1996. "Species, Rules and Meaning: The Politics of Language and the Ends of Definitions in 19th Century Natural History." *Studies in History and Philosophy of Science*, 27(4): 473–519.

Medin, D. L., and S. Atran. 1999. "Introduction," in *Folkbiology*, ed. D. Medin and S. Atran, 1–15. Cambridge, MA: MIT Press.

Mill, J. S. 1843. *A System of Logic, Ratiocinative and Inductive: Being a Connected View of the Principles of Evidence and the Methods of Scientific Investigation*. London: John W. Parker.

1858. *A System of Logic, Ratiocinative and Inductive: Being a Connected View of the Principles of Evidence and the Methods of Scientific Investigation*, 3rd ed. New York: Harper & Brothers.

1882. *A System of Logic, Ratiocinative and Inductive: Being a Connected View of the Principles of Evidence and the Methods of Scientific Investigation*, 8th ed. New York: Harper & Brothers.

Minelli, A. 1993. *Biological Systematics*. London; Chapman & Hall.

Mishler, B. D., and R. Brandon. 1987. "Individuality, Pluralism, and the Phylogenetic Species Concept." *Biology and Philosophy*, 2: 397–414.

Mishler, B. D., and M. Donoghue. 1992. "Species Concepts: A Case for Pluralism," *The Units of Evolution: Essays on the Nature of Species*, ed. M. Ereshefsky, 121–137. Cambridge, MA: Bradford Books.

Morgan, G. J., and W. B. Pitts 2008. "Evolution Without Species: The Case of Mosaic Bacteriophages." *British Journal of Philosophical Science*, 59: 745–765.

Nelson, K. 1973. "Structure and Strategy in Learning to Talk." *Monographs of the Society for Research in Child Development*, 38: 1–135.

Nelson, G., and N. Platnick 1981. *Systematics and Biogeography: Cladistics and Vicariance*. New York: Columbia University Press.

Niklas, K. J. 1997. *The Evolutionary Biology of Plants*. Chicago: University of Chicago Press.

Normore, C. G. 1990. "Ockham on Mental Language," in *Historical Foundations of Cognitive Science*, ed. J. C. Smith, 53–70. Dordrecht, the Netherlands: Kluwer Academic Publishers.

Ogilvie, B. W. 2006. *The Science of Describing: Natural History in Renaissance Europe*. Chicago: University of Chicago Press.

Okasha, S. 2006. *Species Selection, Clade Selection and Macroevolution*. Oxford: Oxford University Press.

O'Malley, M. 2010. "Ernst Mayr, the Tree of Life and Philosophy of Biology." *Biology and Philosophy*, 25: 529–552.

Ospovat, D. 1981. *The Development of Darwin's Theory: Natural History, Natural Theology and Natural Selection 1838–1859*. Cambridge: Cambridge University of Press.

Panchen, A. L. 1992. *Classification, Evolution, and the Nature of Biology*. Cambridge: Cambridge University Press.

Paterson, H. E. H. 1993. *Evolution and the Recognition Concept of Species*. Baltimore, MD: The Johns Hopkins University Press.

Peters, F. E. 1967. *Greek Philosophical Terms: A Historical Lexicon*. New York: New York University Press.

Pietsch, T. W. 2012. *Trees of Life: A Visual History of Evolution*. Baltimore, MD: The Johns Hopkins University Press.

Plato. 1961. *The Collected Dialogues of Plato*, ed. E. Hamilton and H. Cairns. Princeton: Princeton University Press.

Pradeu, T. 2013. "Immunity and the Emergence of Individuality," in *From Groups to Individuals: Evolution and Emerging Individuality*, ed. F. Bouchard and P. Huneman, 77–96. Cambridge, MA: MIT Press.

Putnam, H. 1973. "Meaning and Reference." *The Journal of Philosophy*, 70(19): 699–711.

Quine, W. V. O. 1976. *The Ways of Paradox and Other Essays*. Cambridge, MA: Harvard University Press.

Ray, J. 1735. *The Wisdom of God Manifested in the Works of Creation*, 10th ed. London: William Innys and Richard Manby.

Richards, R. A. 2003. "Character Individuation in Phylogenetic Inference." *Philosophy of Science*, 70: 264–279.

2005. "Evolutionary Naturalism and the Logical Structure of Valuation: The Other Side of Error Theory." *Cosmos and History: The Journal of Natural and Social Philosophy*, 1(2): 270–292.

2008. "Functional Analysis and Character Transformation," in *Form and Function in Developmental Evolution*, ed. M. D. Laubichler and J. Maienshein. Cambridge: Cambridge University Press.

2010. *The Species Problem*. Cambridge: Cambridge University Press.

Ruse, M. 1987. "Biological Species: Natural Kinds, Individuals or What?" *The British Journal for the Philosophy of Science*, 38(2): 225–242.

1999. *The Darwinian Revolution*. Chicago: University of Chicago Press.

Simpson, G. G. 1961. *Principles of Animal Taxonomy*. New York: Columbia University Press.

Slater, M. 2013. *Are Species Real: An Essay on the Metaphysics of Species*. New York: Palgrave Macmillan.

Sloan, P. R. 1976. "The Buffon-Linnaeus Controversy." *Isis*, 67(3): 356–375.

1979. "Buffon, German Biology, and the Historical Interpretation of Biological Species." *The British Journal for the History of Science*, 12(2): 109–153.

1985. "From Logical Universals to Historical Individuals." *Histoire du Concept D'Espece dans Les Sciences de la Vie*. Paris: Fondation Singer-Polignac.

2003. "Reflections on the Species Problem," in *The Philosophy of Marjorie Grene*, ed. R. Auxier and L. E. Hahn. Chicago: Open Court Publishing.

Sober, E. 1988. *Reconstructing the Past: Parsimony, Evolution and Inference*. Cambridge, MA: MIT Press.

2000. *Philosophy of Biology*, 2nd ed. Boulder, CO: Westview Press.

Sokal, R. R., and P. H. A. Sneath 1963. *Principles of Numerical Taxonomy*. San Francisco: W. H. Freeman and Company.

Sorabji, R. 1990. "The Ancient Commentators on Aristotle," in *Aristotle Transformed*, ed. R. Sorabji, 1–30. Ithaca, NY: Cornell University Press.

Stafleu, F. A. 1971. *Linnaeus and the Linnaeans*. Utrecht, the Netherlands: A. Oosthooek's Uitgeversmaatschappij N.V.

Stamos, D. 2003. *The Species Problem*. Lanham, MD: Lexington Books.

2007. *Darwin and the Nature of Species*. Albany: State University of New York Press.

Stauffer, R. C. 1975. *Charles Darwin's Natural Selection*. Cambridge: Cambridge University Press.

Stevens, P. F. 1983. "Augustin Augier's "Arbre Botanique" (1801), A Remarkably Early Botanical Representation of the Natural System." *Taxon*, 32(2): 203–211.

1994. *The Development of Biological Systematics: Antoine-Laurent de Jussieu, Nature and the Natural System*. New York: Columbia University Press.

Stott, R. 2003. *Darwin and the Barnacle*. New York: W. W. Norton.

Thompson, A. 1995. "The Debate on Universals before Peter Abelard." *The Journal of the History of Philosophy*, 33(3): 409–429.

Van Valen, L. 1992. "Ecological Species, Multispecies, and Oaks," in *The Units of Evolution: Essays on the Nature of Species*, ed. M. Ereshefsky, 69–77. Cambridge, MA: Bradford Books.

Velasco, J. 2012. "The Future of Systematics: Tree Thinking Without the Tree." *Philosophy of Science*, 79(5): 624–636.

Waxman, S. R. 1999. "The Dubbing Ceremony Revisited: Object Naming and Categorization in Infancy and Early Childhood," in *Folkbiology*, ed. D. L. Medin and S. Atran, 233–284. Cambridge, MA: MIT Press.

Weishampel, D. B., P. Dodson, and H. Osmólska. 2004. *The Dinosauria*, 2nd ed. Berkeley: University of California Press.

Wheeler, W. 2012. *Systematics: A Course of Lectures*. Oxford: John Wiley & Sons.

Whewell, W. 1989. *Theory of Scientific Method*, ed. R. E. Butts. Indianapolis, IN: Hackett.

Wiley, E. O. 1981. *Phylogenetics: The Theory and Practice of Phylogenetic Systematics*. New York: John Wiley & Sons.

Wiley, E. O., and B. S. Lieberman, 2011. *Phylogenetics: The Theory of Phylogenetic Systematics*. Hoboken, NJ: John Wiley & Sons.

Wilkins, J. 2009. *Species: A History of the Idea*. Berkeley: University of California Press.

Wilkins J. S., and M. C. Ebach, 2014. *The Nature of Classification: Relationships and Kinds in the Natural Science*. London: Palgrave Macmillan.

Wilson, E. O. 1999. *The Diversity of Life*. London: W. W. Norton.

2000. *Sociobiology: The New Synthesis*. Cambridge, MA: Harvard University Press.

Winsor, M. P. 1969. "Barnacle Larvae in the Nineteenth Century: A Case Study in Taxonomic Theory." *Journal of the History of Medicine and Allied Sciences*, 24: 294–309.

2001. "Cain on Linnaeus: The Scientist-Historian as Unanalysed Entity." *Studies in History and Philosophy of Biology & Biomedical Science*, 32(2): 239–254.
2003. "Non-essentialist Methods in pre-Darwinian Taxonomy." *Biology and Philosophy*, 18: 387–400.
2006. "The Creation of the Essentialism Story: An Exercise in Metahistory." *History and Philosophy of Life Sciences*, 28: 149–174.
Wittgenstein, L. 1968. *Philosophical Investigations*. New York: Macmillan.

Index

Printed in the United States
By Bookmasters